THE BIOLOGY OF DESIRE

the

BIOLOGY *of*

DESIRE

Why Addiction Is Not a Disease

MARC LEWIS

PUBLICAFFAIRS
New York

Published in the United States by PublicAffairs™,
A Member of the Perseus Books Group

PublicAffairs books are available at special discounts for bulk purchases in the US
by corporations, institutions, and other organizations. For more information, please
contact the Special Markets Department at the Perseus Books Group, 2300 Chestnut
Street, Suite 200, Philadelphia, PA 19103, call (800) 810-4145, ext. 5000, or e-mail
special.markets@perseusbooks.com.

Book design by Jeff Williams

A CIP catalog record is available for this book from the Library of Congress.

ISBN 978-1-61039-437-6 (hardcover)
ISBN 978-1-61039-438-3 (e-book)
Library of Congress Control Number: 2015940383

First Edition

10 9 8 7 6 5 4 3 2

For the members of my blog community, who have
generously shared their experiences and insights,
and for the five who trusted me to tell their stories here

CONTENTS

INTRODUCTION

Public attention has been riveted by the harm addicts cause themselves and those around them, more in the last few years than ever before. And the way we view addiction is changing, moulting, and perhaps advancing at the same time. We've begun to separate our ideas about addiction from assumptions about moral failings. We're less likely to dismiss addicts as simply indulgent, spineless, lacking in willpower. It becomes harder to relegate addiction to the down-and-outers, the gaunt-faced youths who shuffle toward our cars at traffic lights. We see that addiction can spring up in anyone's backyard. It attacks our politicians, our entertainers, our relatives, and often ourselves. It's become ubiquitous, expectable, like air pollution and cancer.

To explain addiction seems more important than ever before. And the first explanation that occurs to most people is that addiction is a disease. What else but a disease could strike anyone at any time, robbing them of their well-being, their self-control, and even their lives? Many esteemed public health organizations and doctors call it a disease. Rehabs, addiction counsellors, and twelve-step fellowships call it a disease. Research over the last twenty years has found indisputable evidence for changes in brain structure and function that parallel substance abuse. And genetic studies reveal heritable traits that predispose people to addiction. All this seems to clinch the definition of addiction as a disease—a physical disease. And that gives us hope, or at least forbearance, because the notion is sensible, comforting in its own way, and part of our shared reality. If addiction is a disease, then it should have a cause, a time course, and possibly a cure, or at least agreed-on methods of treatment. Which means we can hand it over to the professionals and follow their instructions.

But is addiction really a disease?

This book makes the case that it isn't. Addiction results, rather, from the motivated repetition of the same thoughts and behaviours until they become habitual. Thus, addiction develops—it's learned—but it's learned more deeply and often more quickly than most other habits, due to a narrowing tunnel of attention and attraction. A close look at the brain highlights the role of desire in this process. The neural circuitry of desire governs anticipation, focused attention, and behaviour. So the most attractive goals will be pursued repeatedly, while other goals lose their appeal, and that *repetition* (rather than the drugs, booze, or gambling) will change the brain's wiring. As with other developing habits, this process is grounded in a neurochemical feedback loop that's present in all normal brains. But it cycles more persistently because of the frequent recurrence of desire and the shrinking range of *what* is desired. Addiction arises from the same feelings that bind lovers to each other and children to their parents. And it builds on the same cognitive mechanisms that get us to value short-term gains over long-term benefits. Addiction is unquestionably destructive, yet it is also uncannily normal: an inevitable feature of the basic human design. That's what makes it so difficult to grasp—socially, scientifically, and clinically.

I believe that the disease idea is wrong, and that its wrongness is compounded by a biased view of the neural data—and by doctors' and scientists' habit of ignoring the personal. It's an idea that can be replaced, not by shunning the biology of addiction but by examining it more closely, and then connecting it back to lived experience. Medical researchers are correct that the brain changes with addiction. But the *way* it changes has to do with learning and development—not disease. Addiction can therefore be seen as a developmental cascade, often foreshadowed by difficulties in childhood, always boosted by the narrowing of perspective with recurrent cycles of acquisition and loss. Like other developmental outcomes, addiction isn't easy to reverse, because it rides on the restructuring of the brain. Like other developmental outcomes, it arises from neural plasticity, but its net effect is a reduction of further plasticity, at least for a while. Addiction is a habit, which, like many other habits, gets entrenched through a decrease in self-control. Addiction is definitely bad news for the addict and all those within range. But the severe consequences of addiction don't make it a disease, any more than the consequences of violence make violence a disease, or the consequences of racism make racism a disease, or the folly of loving thy neighbour's wife makes infidelity a disease. What they make it is a very bad habit.

Although this book uses scientific findings to build its case, it works through the testimony of ordinary people. I relate detailed biographical narratives of five very different people, each struggling with addiction, as the scaffolding on which brain science is introduced and interpreted. I have rendered these narratives in a literary style, including stream of consciousness and dialogue, but they are factually accurate, except for the use of pseudonyms and the inexact wording of some of the dialogue. Through these stories, I show what it's like and how it feels when addiction takes hold, while explaining the neural changes underlying it. There's no doubt that these changes mark a difficult passage in personality development. But I conclude each chapter on a positive note, following my contributors through their addictions to their growth beyond it—a phase often termed "recovery." And I provide the neuroscientific facts and concepts to help us understand how they get there. The many addicts who end up quitting do so uniquely and inventively, through effort and insight. Thus quitting is best seen as further development, not "recovery" from a disease.

∽

I'm a neuroscientist and a professor. It's my job to teach students whatever I know about the brain. I've taught and done research on emotional development and the brain for most of my career. But after my first decade of lecturing, I began to sound stodgy and dull, even to myself. What was I missing? The brain is the foundation of our needs, our desires, our joy and suffering, our darkest moments and our capacity to overcome them. Why was it coming across as an anatomical jigsaw puzzle, a blueprint for a circuit board, a thicket of labels, boxes, and arrows? How could I convey the gut-wrenching reality of the brain as a motivational furnace? Even in graduate courses, students met my efforts with glassy stares and furious note taking. Look up! I wanted to shout. Look up from your notes and feel what your brain is doing. You can get this directly. Not from your notes. Just introspect a bit and you'll discover that your brain is busily extending and revising a landscape of flitting thoughts, shocking associations, and childish impulses. It's not just an organ of rationality, as you've no doubt been taught; it's also the biological engine of our striking irrationality—it has a dark side. How does *that* work?

And how do I get it across?

About six or seven years ago I began to talk more candidly about my own messy emotions. I culled examples from my past, exposing the dark side of my own brain. That got their attention.

Especially when I revealed that I'd been a drug addict through most of my twenties—something I'd locked away from public scrutiny for nearly thirty years. Professors aren't supposed to be drug addicts, past, present, or future. This was interesting. At around the same time, I began riffling through the journals I'd kept from my late teens to mid-thirties. I relived hundreds of traumatic, horrific, and often baffling experiences of getting high and getting lost. I began to read and think about the brain processes underlying addiction, and I began the book I hoped would put it all together: my previous book, *Memoirs of an Addicted Brain*.

I stopped taking illegal drugs and taking drugs illegally at the age of thirty. Now, as a neuroscientist and a teacher, I needed to figure out what had happened to me all those years ago. How had my brain become so addled for such a long time? How did I finally quit? As I waded through a sea of papers on the neuroscience of addiction, I learned how circuits devoted to goal seeking become captivated by the appeal of a single goal. A drug, a drink, gambling, porn—whatever it is that satisfies a powerful desire, at least partially, while simultaneously increasing its own appeal. I started to understand the dark side of the brain as a scientist as well as an "end user"—and I began to convey what I was learning to my students, with passion, precision, and, I hope, insight.

This book is my current attempt to be that teacher. While I have a message to get across, an argument to make about addiction, my most daunting task is to move back and forth between two perspectives: life as we experience it—including its pinnacles and perils—and the concrete workings of the brain that make that experience possible. If we are to understand anything so complex and troubling as addiction, we need to gaze directly at the point where experience and biology meet. Because that's the bottleneck, the linchpin, where human affairs are cast and crystallized. That's where the brain shapes our lives and our lives shape the brain.

ONE

Defining Addiction

A Battleground of Opinions

O ver the past few decades, society has come to see addiction as a specific, definable phenomenon, rather than some moral deficit or personal fall from grace. However, there is little consensus on how to conceptualize this phenomenon. In our efforts to study the nature of addiction, delineate its causes, and explore potential treatment strategies, we have come up with a variety of (mostly incompatible) definitions. These can be narrowed down to three broad categories: disease, choice, and self-medication.

According to the disease model in its current form, addiction is a brain disease. It is characterized by changes in specific brain systems, especially those that process rewards (i.e., valued outcomes). Brain systems responsible for anticipating rewards, motivating us to go after them, and evaluating and reevaluating the worth of those rewards are reshaped by the repeated use of drugs, including alcohol. Researchers have found additional brain changes in systems underlying cognitive control, delayed gratification, and abstract skills like comparing and predicting outcomes and selecting best choices. According to the disease model, all these changes are caused by exposure to drugs of abuse, and they are difficult if not impossible to reverse. Of course the disease model builds on a biological framework, and it does a good job of explaining why some individuals are more vulnerable to addiction than others, based on genetic differences and other dispositional factors. And the cure? Well, there doesn't seem to be one. Addiction is currently viewed as a chronic disease. But that's not a problem for the disease model, because it's also true of many well-known illnesses, including heart disease, diabetes, and some forms of cancer. For those too there are treatments but not cures.

The idea that addiction is a choice comes from a cognitive (rather than biological) perspective, emphasizing changes in thought processes. Researchers in behavioural economics, which blends social psychology with economic thinking, try to understand why people make the choices they make, including the choice to take addictive substances. While few people imagine that addiction is a good choice, it is often considered a rational one, at least in the short run—as when the pleasure or relief derived from one's addiction seems to outweigh other possible choices. Unfortunately, the choice model provides a convenient platform for those who consider addicts indulgent and selfish. If addiction is a choice, they reason, then addicts are deliberately inflicting harm on themselves and, more seriously, on others. Yet other proponents of the choice model point to environmental or economic conditions beyond the addict's control, including poverty and social isolation. The choice model does a better job than the disease model of explaining how addicts quit. When conditions change with time and circumstances, so do choices. It should not be surprising, then, if people choose to quit when life circumstances improve, or when the financial or social costs of remaining addicted exceed the benefits of being high. Either explanation could account for the undisputed finding that a majority of heroin-addicted veterans stopped using heroin when they returned from Vietnam.

The self-medication model is a hodgepodge. It derives partly from psychology, partly from medicine, and partly from sociology, but it is grounded in developmental thinking. As children and adolescents develop, emotional problems can erode their sense of well-being. They try different strategies to deal with those problems, until they find something that works. Trauma—whether social, psychological, or sexual—is a buzzword for early adversity, and post-traumatic stress disorder (PTSD) is often found to underlie anxiety and depression. Substance abuse among those with PTSD is as high as 60–80 percent, and the rate of PTSD among substance abusers is 40–60 percent—reason enough to believe that people take drugs to relieve stress.[1] In fact, psychoactive drugs are well known to relieve anxiety, interrupt rumination, or brighten one's mood. Whether self-medication is considered a choice or a lucky (at first) accident doesn't really matter. The point is that drugging and drinking make you feel better. Until they don't. A nasty side effect of addictive drugs is that the addiction itself becomes a source of stress—often *the* major source of stress—especially when tolerance is going up, your bank balance is going down, and withdrawal symptoms set in. But that doesn't mean the self-medication model is wrong. Most medicinal regimes have some unpleasant side effects.

These three models of addiction overlap to some degree, but each has unique implications for research, funding, and care, from the level of government policy to that of treatment options for individual sufferers. To put it simply, the disease model calls for treatment at the hands of experts—generally medical experts (including psychiatrists) but also the burgeoning band of treatment personnel who report to them (at least in theory); the choice model advocates reviewing one's beliefs and changing one's perspective, often using standard psychotherapeutic techniques such as cognitive behavioural therapy and motivational interviewing; and the self-medication model stresses the need to protect children and adolescents from extreme psychosocial pressures and to diagnose and treat underlying developmental issues that may have predisposed a person to addiction.

All of these models make some sense, yet none of them, either alone or in combination, has yielded definitive explanations as to how addiction works or how it can be effectively treated. Research on the cause and treatment of addiction absorbs billions of dollars each year, without a great deal of success. We need to understand addiction a lot better if we want those dollars to count. We need to address the central questions that anyone touched by addiction wants answered: What is it? How does it work? Why is it so hard to stop? To answer these questions, we need to blow past the war of definitions and arrive at a coherent, comprehensive model.

THE FRONT-RUNNER

After years of talking, writing, and blogging about addiction, I get a lot of emails from addicts: some still at the height of their addiction, some recovering, some well past it. And every so often, one of those emails really moves me. Here's a passage from one that came just two weeks before I started writing this book, from a woman I've never met and probably never will—a meth addict:

> I am unsure of what to do or where to turn next. I tried rehab once
> for a few days before my body became toxic and I ended up in the
> hospital for a week. It was only after I tried quitting that i fell ill,
> close to death with a high fever, failing kidneys, and toxemia. Now
> three years later I am that much more addicted and afraid that this is
> what will kill me, and it won't be long. I don't know what I am more

afraid of, being sick physically and dying or staying high, falling apart mentally, and for things to never change. Maybe this is how it was meant to be? In which case life isn't worth living and my children might be better off without me. I wish there was an antidote.

Not only is she suffering terribly, but her suffering seems to be without any purpose—and completely beyond her control. That's what addiction can feel like and what it can look like. And that's why most people see addiction as a disease. The disease model is clearly the front-runner among current attempts to define addiction.[2] The idea that addiction is a disease is accepted almost everywhere, and most addicts have little choice but to go along with this definition and submit to the treatment policies that derive from it. But the loss of control experienced by addicts and their families is only one reason for treating addiction as a disease. Others come from researchers and clinicians themselves.

As medical science becomes more sophisticated, its doctrines and policies also become more persuasive, more difficult to ignore. Doctors, medical researchers, and health policy makers assume that addiction is a disease because, for one thing, that's how medicine defines human problems. For another thing, the multimillion--dollar price tags of sophisticated research programs are funded by mainstream medical institutions, like the National Institutes of Health (NIH). So it's no accident that the flashiest research maintains an intimate connection with the medical mainstream. Which is not to say that the research findings haven't been impressive. They have. It's just that it's many times easier to perform impressive research on the workings of cells than the workings of families or cities—because that's where the money is, and because cells are easier to observe. But there's a third reason to back the disease model: it does a good job of enfolding the other two. You can say you're making a choice. But if it's a really bad choice, an obviously bad choice, an illogical, stupid, self-destructive choice, then one can argue that your choice-making mechanism is, well, diseased. And you might have gotten to that state trying to put out the emotional fires that sprang up when you were fourteen—along the lines of the self-medication model. Yet putting bad things in your body, for whatever reason, is known to trigger long-term physiological consequences—such as disease.

The fact is that we in the West embrace the logic of pigeonholing problems, giving them unique names, and finding technical solutions—the more targeted the better—for alleviating them. That is, to a T, the logic of Western medicine.

Here are the specifics. According to the National Institute on Drug Abuse (NIDA—a component of NIH), "Addiction is defined as a chronic, relapsing brain disease that is characterized by compulsive drug seeking and use, despite harmful consequences."[3] Specifically, the metabolism of dopamine, a crucial neurotransmitter for motivating and directing goal-seeking behaviour, is altered. So, over time, only the user's substance of choice is capable of triggering dopamine release (or reception) in the brain regions responsible for motivation and meaning. This is the accepted—in fact, nearly unchallenged—stand taken by the medical community, the psychiatric community, and the addiction research community, as reflected by a mountain of articles and posts by NIDA, NIH, the American Medical Association (AMA), and the American Society of Addiction Medicine (ASAM). According to Steven Hyman, MD, previous director of the National Institute of Mental Health (NIMH), addiction is a condition that changes the way the brain works, just like diabetes changes the way the pancreas works. Hyman goes on to argue that "in vulnerable individuals, the disease of addiction is produced by chronic administration of the drugs themselves."[4] In other words, by taking drugs, you're making yourself sick.

Nora Volkow, the present director of NIDA, esteemed neuro-scientist, and firebrand orator, appears highly confident about the science underlying the disease model. In one NIDA publication she declared: "As a result of scientific research, we know that addiction is a disease that affects both the brain and behavior. We have identified many of the biological and environmental factors and are beginning to search for the genetic variations that contribute to the development and progression of the disease."[5] More recently, NIDA's position was elaborated further: "Brain imaging studies of people with addiction show physical changes in areas of the brain that are critical to judgment, decision making, learning and memory, and behavior control."[6]

In fact, Volkow points to changes in the structure and function of the brain as indisputable support for the disease model. Not only the brain regions underlying impulsive goal seeking, says Volkow, but also those responsible for self-control are physically damaged by drugs and can no longer function normally. Motivated action—the wish to do, to go, to acquire—comes under the influence of drugs or booze, and the normal pleasures of life cease to register. Meanwhile, the cortical safeguards designed by evolution to control such preoccupations are three floors up in intensive care.

The brain disease model is supported by two pillars that have proven rather difficult to crack. The first is the corpus of evidence that the brain really does change with addiction. That's a stumper for advocates of the choice model. Once people become aware that addiction changes the brain, it makes sense to define it as a brain disease. After all, the pancreas changes with diabetes, and the heart changes with heart disease. The liver changes with hepatitis, and the lungs change with lung cancer. If a condition changes the shape or function of our organs, and if that change is difficult or impossible to reverse, we label that condition a disease. The other pillar is the control issue. Addicts really do seem to have lost control. At least that's what they say, and that's how it looks. Loss of control is about the worst thing that can befall anyone. From age three to age ninety-three, being in control is synonymous with being okay. In fact, the word "disease" and its close cousin "disorder" call up images of a system that's out of whack, that behaves erratically, that is essentially out of control, like a car with defective steering. And biological diseases can be characterized as structural or functional failings that cause the body to wobble, like a gyroscope losing its equilibrium, with the likely outcome of falling over.

I recently met Nora Volkow at a weeklong "dialogue" with the Dalai Lama—an initiative organized by the Mind and Life Institute, a nonprofit organization committed to building a scientific under- standing of the mind. She and I were among eight presenters on the theme of "Craving, Desire, and Addiction." I was familiar with Dr. Volkow's research and policy initiatives, but I wanted to make sure I wasn't missing anything. It seemed I wasn't. We exchanged volleys over a lunch break. I had challenged the disease model during a recent discussion period, emphasizing that brain change could be interpreted as a consequence of learning rather than dis- ease. What did she think of that argument? She told me exactly what she thought: that I was wrong. Cocaine has been shown to damage the brains of rats, she said, just as alcohol (in sufficient doses) dam- ages our own brains. Furthermore, when something doesn't work the way it should, we call it a disease. That's how the word is used, she insisted. Yet her main objective wasn't a war of definitions; it was to facilitate treatment for the people who needed it. Calling ad- diction a disease not only mitigates massive volumes of stigma and guilt but also aims to provide accessible avenues for addicts to get help. This seemed to be her bottom line. Diseases require interven- tion, and even the leaky health care system in the United States is inclined to try to fight the disease of addiction.

I liked this ardent, energetic woman from the moment I met her. She speaks without hesitation, with a spiky self-assurance gleaned from her faith in scientific methodology. She defends her position with a data-stuffed cranium. And she is persuasive! But I couldn't buy it. The upshot seemed to be that if we don't treat addiction as a disease, it won't be treated at all. In a country where the social safety net is itself a fragile concept, where child poverty is second-highest among all developed countries,[7] perhaps the disease model *is* a useful means for getting care to those who need it or, more likely, needed it a decade ago. But that doesn't make it good science. And bad science makes for models of treatment that are distorted and ineffective.

Volkow and other experts see the disease model as an improvement over the centuries-old norm of denigrating addicts for their lack of willpower and moral decrepitude. And that's certainly a step in the right direction. Despite the despicable things addicts sometimes do, intense shame and guilt are more likely to thwart recovery than facilitate it. After all, these are painful emotions, and most addictive substances and acts provide some form of pain relief. The disease model has also served to stimulate volumes of new research, promote the development of useful medications, and consolidate our understanding that addiction involves biological factors.

Yet, after reading thousands of comments and emails from former and recovering addicts and interviewing dozens of them for hours at a time, I'm convinced that calling addiction a disease is not only inaccurate, it's often harmful. Harmful, first of all, to addicts themselves. While shame and guilt may be softened by the disease definition, many addicts simply don't see themselves as ill, and being coerced into an admission that they have a disease can undermine other—sometimes highly valuable—elements of their self-image and self-esteem. Many recovering addicts find it better *not* to see themselves as helpless victims of a disease, and objective accounts of recovery and relapse suggest they might be right. Treatment experts and addiction counsellors often identify empowerment or self-efficacy as a necessary resource for lasting recovery. And, in a statistically rigorous but highly provocative study, Miller and colleagues found that the *only* pretreatment characteristic that predicted relapse, six months after concluding outpatient treatment for alcohol dependence, was "the extent to which clients endorsed disease model beliefs before entering treatment."[8] As well, once they've beaten their addictions, many former addicts choose not to spend the rest of their days walking on eggshells lest the disease return. While re-addiction is clearly a hazard for some, others achieve a realistic and lasting confidence that they've outgrown their addictions and it's time to move on. In fact, survey research published over the last thirty years indicates that most addicts eventually recover permanently.[9] For them, the disease label may be an unnecessary, even harmful, burden.

Most of the recovered addicts I've talked to would rather think of themselves as free—not cured, not in remission. Having overcome their addictions by dint of hard work, intense self-examination, and the courage and capacity to regrow their perspectives (and their synapses), they'd rather see themselves as having *developed* through addiction and become stronger as a result. Neuroscientific findings actually support this intuition—once neuroscience steps away from the funding priorities set by the medical mainstream (e.g., NIDA). And that's another reason I'm writing this book: to give addicts what they need, and what they deserve, by interpreting the scientific data in a way that actually corresponds with their experience of what they've been through and their sense of who they are. Science, meet subjective experience. Subjective experience, meet science. I'd like you two to try to get along.

THE RISE AND (ANTICIPATED) FALL
OF THE DISEASE MODEL

The idea that addiction is some kind of disease is unquestionably the dominant view in government, medical, and most scientific circles around the world. So dominant in the West, for example, that US vice president Joe Biden introduced the Recognizing Addiction as a Disease Act for debate in the US Senate on March 28, 2007.

S. 1011: Recognizing Addiction as a Disease Act of 2007

(1) Addiction is a chronic, relapsing brain disease that is characterized by compulsive drug seeking and use, despite harmful consequences. It is considered a brain disease because drugs change the brain's structure and manner in which it functions. These brain changes can be long lasting, and can lead to the harmful behaviors seen in people who abuse drugs. The disease of addiction affects both brain and behavior, and scientists have identified many of the biological and environmental factors that contribute to the development and progression of the disease.[10]

Yet the concept that addiction is a disease is certainly not new. In fact, it's been promoted and rebutted since the time of Aristotle (and other Greek and Egyptian scholars), and it has grown exponentially in authority and popularity since the early 1900s.[11] This quote from a hundred years ago captures the flavour of the disease concept when it began to proliferate in the West:

The author considers it very unfortunate that the terms "morphine habit" and "opium habit" have been, and are still, so universally employed when referring to narcotic addiction (disease). They are misleading and do not, in any wise, accurately describe the condition present. . . . Habit implies something that can be corrected by an exercise of the will. . . . This is not true of narcotic disease; therefore, it is not a mere habit and should not be spoken of as such. . . .

The man who is addicted to a narcotic drug is as truly a diseased man as one who has typhoid fever or pneumonia.[12]

How did this definition arise, and how has it evolved in our own time?

Society's conceptualization of addiction has always reflected its policies for dealing with it. While Shakespeare referred to addiction in *Henry V,* nobody at the time advocated treatment centres for debauched nobility. Public alarm began to rise over "demon rum" and other spirits early in the nineteenth century. By the end of that century, temperance movements vociferously demanded total abstinence. In the early twentieth century, alcoholics and addicts were seen as both doomed and damned if they could not or would not dry up. And when those warnings didn't work, Prohibition was launched in the United States as the next-best solution. Temperance activists saw alcohol itself as the cause of alcoholism, much like contemporary disease theorists see drugs (rather than environments) as the cause of drug addiction. When Prohibition was repealed to allow for social drinking, hard-core alcoholics were nevertheless reviled as morally depraved and undeserving of help.

Public policy thus maintained its moralistic and puritanical slant well into the 1930s. But then the view of medical practitioners, that addiction was a malady rather than a personal failing, picked up support from an unexpected source. Bill Wilson and Robert Smith started Alcoholics Anonymous (AA) in 1935, launching a new era in society's perception of addicts and their treatment. The premise of AA was that alcoholics were suffering human beings who had the right and the obligation to try to relieve their suffering. Through principles of mutual support, ongoing group attendance, self-honesty, and spiritual transformation, AA helped millions of alcoholics overcome their addictions, as it still does today. It also spearheaded society's recognition that addicts need help, not rejection, and that they *can* get better.

The founders of AA did not see addiction as a disease, exactly, but as a mental and spiritual "malady." Physical sensitivity to alcohol was initially conceived of as an "allergy," while the spiritual malady expressed itself in perpetual discomfort with life on life's terms, an inability to be at peace in the moment. Booze first seemed to soothe this discomfort but ultimately exacerbated the physical sensitivity. The result was a lifelong disorder that remained treatable, though never actually curable. AA counselled its members to stay vigilant about their vulnerability and to keep it firmly in mind by reciting metaphors, chanting slogans, and telling and retelling personal tales of failure and of success. A crucial springboard to sobriety was realizing that you were powerless over alcohol—you were not capable of moderate or occasional drinking. The twelve steps of AA begin, still today, with an admission of powerlessness and a commitment to trust in a higher authority. It turns out not to matter as much anymore whether that authority is God, the group itself, your sponsor, or the medical community. What does matter is the acknowledgement of a serious deficit, which is—not coincidentally—the state you find yourself in when the doctor says you have cancer or pneumonia. That's when you know you need help.

While AA's emphasis was on the mental and spiritual aspects of addiction, the idea of a biological sensitivity to alcohol opened the door to a more specific (and broadly accepted) definition of addiction as a disease. In the early 1950s, when Narcotics Anonymous (NA) and Hazelden's "Minnesota Model" got off the ground, the disease nomenclature began to flourish. NA, an outgrowth of AA, was established to treat those addicted to heroin and other drugs, and it was considered self-evident that the drug was what caused the disease of addiction. The Minnesota Model, which blended twelve-step philosophy with principles of residential care and education, became the gold standard for treatment centres by the 1960s. The Minnesota Model specifically labelled alcoholism a disease that overcame people physically, mentally, and spiritually. At the same time, an influential book by E. M. Jellinek, *The Disease Concept of Alcoholism,* articulated a medical model that traced the progression of alcoholism through a series of phases leading to loss of control, insanity, and death. Now the "disease" terminology began to appear in the literature of twelve-step programs throughout North America. And the American Medical Association classified alcoholism as an "illness" in 1967, making it official. In retrospect, the concept of a biological deficit, reified by AA, helped pave the way for the disease concept of addiction, and a medical term became standard parlance in the world of addiction treatment.

Today the disease definition is used by twelve-step programs around the world, though its meaning continues to morph and vary from group to group. Moreover, twelve-step methods have become central in the world of institutional treatment, where the disease definition has been imported wholesale. There is a basic incompatibility between AA philosophy and the impersonal character of institutional care, and the disease label just reinforces the resulting fallout. Addicts seeking treatment, or those coerced into treatment by the justice system, are compelled to follow a recipe for recovery targeted to what is viewed as their disease, independent of their personal beliefs, which are often dismissed as irrelevant. If they do not follow the recipe, they may be denied any treatment at all, a policy that is fundamentally at odds with official twelve-step literature (though some twelve-step groups adopt the same punitive methods). For many addicts, this pressure tactic is a deal breaker, and that helps explain the acrimonious tone of the criticisms expressed by those who've quit or been excluded from twelve-step-based care.

There are other ways in which twelve-step practice has helped erect barriers while attempting to relieve suffering. First, the AA framework and the medical notion of disease share the core assumption that addiction is a lifelong disorder and total abstinence is necessary to arrest it. The graded (e.g., occasional, social) use of any substance is deemed self-destructive, inevitably leading to relapse. This position often strikes former addicts as exaggerated and untenable, and epidemiological research shows that many recovered alcoholics are capable of social drinking. (The debate about moderation versus total abstinence is contentious and volatile, and I won't attempt to get into it here. Suffice it to say that many sources of evidence point to highly individualistic outcomes and resting points along the route to recovery. And whether or not total abstinence is necessary has little bearing on the disease concept, regardless.) Second, the collaboration between the twelve-step movement and institutional thinking asserts the need for treatment through recognized programs. This policy discourages addicts from finding their own way to recovery, and it blocks their access to benefits that might help pay for alternative resources. Moreover, it ignores compelling data, collected by a variety of independent organizations (most famously the National Epidemiologic Survey on Alcohol and Related Conditions, NESARC), showing that most addicts and alcoholics do recover, and that a majority of those—up to three-quarters, depending on where you get your statistics—recover without *any* treatment. Third, twelve-step literature maintains that the disease of addiction is built into one's character. Experts including Stanton Peele have shown how destructive this attribution can be, especially for young people whose identities are still under construction.

Finally, and most troubling, is the confusion that surrounds AA's emphasis on recognizing one's "powerlessness" as a condition for overcoming addiction. For those helped by twelve-step methods, powerlessness is usually viewed as a hinge point for surrendering unworkable strategies and admitting that one has to start over and revamp one's design for quitting. However, others interpret the emphasis on powerlessness as suggesting ongoing helplessness, perhaps because their thinking has been distorted by submission to a set of impersonal rules imposed by the courts, institutional policies, or overly severe group leaders. As I noted earlier, many experts highlight the value of empowerment for overcoming addiction. In fact, most former addicts claim that empowerment, not powerlessness, was essential to them, especially in the latter stages of their recovery. Sensitivity to the meaning of empowerment in recovery may be greatest for those who've been disempowered in their social world, including women, minorities, the poor, and those with devastating family histories.

It's an open question whether the disease nomenclature, partially absorbed into the AA mainstream, has alienated more members than it's helped. Here's a comment I received about a year ago, following a blog post on the disease label:

I am a Registered Professional Counsellor and I have personally struggled with alcohol addiction in my life.

After the last three years of intense psycho-therapy and group work focused on healing personal wounds from our childhood and dealing with our traumas, I have managed to come out of my addiction on to the other side.

I have many friends who still rely heavily on the AA program, and with no disrespect to the program—I can see how it works for them, it just does not work for me.

I have had a long hard look inside about how I feel personally about addiction. I do not feel that I have or had a disease. I see my past drinking as a behavioral problem, a learned response to dealing (or not dealing) with emotional pain and stress. Once I achieved the excavating of my wounds I no longer lived with the same anxiety or sense of dread/guilt and shame. . . . I have completed the steps—however, I see them as steppingstones rather than a Solution.

The disease concept evolved from a description to a model in the 1990s—"the decade of the brain."[13] Neuroscientists began to show new synaptic growth in morphine-addicted lab rats and neural rewiring in human cocaine addicts: clear evidence of brain change. With other drugs the story was sometimes more complicated, but the fundamental message was the same: drug use messes up brain wiring, and the mess doesn't disappear when you quit. Many of the reported structural changes were related to changes in the release and absorption of dopamine, a neurochemical associated with reward in subcortical systems but with cognitive control in the loftier reaches of the cortex. In study after study, dopamine levels went up and down with drug availability—and not much else. Dopamine was increasingly released by getting high, or by cues that predicted getting high, or by cues that predicted cues that predicted getting high, and decreased in relation to other formerly pleasurable activities like sex, food, and watching your kids grow up. The brain receptors that absorb and use dopamine were also found to change in structure or efficiency over months and years of use.

Because the action of dopamine enhances the formation of new synapses (and the corresponding loss of older ones), changes in the dopamine system bring about structural changes in synaptic networks—the basic wiring diagram of the brain. And they do so most significantly in a brain region called the striatum, the area responsible for pursuing rewards. These brain changes were seen as direct evidence that an insidious force—namely, drugs—had "hijacked the brain," a phrase first uttered by Bill Moyers on a popular PBS television series but quick to catch on in addiction debates everywhere. I'll delve more deeply into brain change in subsequent chapters. For now, what's important to emphasize is the impact of such findings on the conceptualization of addiction, comfortably defined as a "chronic brain disease" from the late 1990s to the present.

It makes sense that medical practitioners (and their colleagues in related professions) readily jumped on the bandwagon. First, it jibed with psychiatrists' long-standing efforts to "medicalize" psychological problems, to see mental illness through a biological lens, so that medical doctors (especially psychiatrists) remained the ruling experts on psychological matters. Second, by fitting addiction within a medical category, the disease model provided coherence and closure in a field customarily sown with discord. Doctors rely on categories to understand people's problems, even problems of the mind. Every mental and emotional problem is identified with a medical label, from borderline personality disorder to autism, depression, anxiety, and addiction. These conditions are described as tightly as possible and listed in the *Diagnostic and Statistical Manual of Mental Disorders* (DSM) and the International Classification of Diseases (ICD). In fact, the DSM is famous for categorizing every nuance of personal disturbance as a type or subtype of disease, and the latest rewrite of the DSM—creatively labelled DSM-5—can be seen as leading to more medicalization because it includes more symptoms. It would be strange indeed if addiction were not invited to join the club.

Since our opinions and convictions are so firmly guided by the dictates of medicine, the disease concept has become a juggernaut, overtaking diverse arenas of public opinion and public health. Thousands of self-help books, websites, and YouTube videos spread the word: *Addiction is nothing to be ashamed of. It's a brain disease.*

As argued earlier, the disease model probably does more harm than good for most addicts. Yet its benefits for other players are clear. The disease model is excellent news for the owners and managers of the more than fifteen thousand drug and alcohol rehab centres operating in the United States and Canada, because it means *We know what your problem is, and we're the ones to fix it.* Drug and alcohol treatment and rehab represents a multibillion-dollar industry in the Western world. (Costs vary from country to country but are generally above $2,500 per week in the United States and Canada, slightly lower in Britain and Europe.) And while the size of the problem may justify the enormous size of this network, we must recognize the industry as a special interest with much to gain and much to lose.

The definition of addiction as a disease, endorsed by the medical and scientific communities and most Western governments, may be the most powerful marketing tool there is for the rehab industry. It's not only a great way to get people in the door—clearly people with a disease need treatment, and judges in the United States have fully endorsed this logic—but also a way of explaining what goes wrong when treatment doesn't work. Because no doctor, nurse, or shrink will ever tell you that they can fix you for sure. All they can say is that they'll try. And if you end up *not* getting fixed, well, that's the way it is with diseases. And probably you didn't quite follow the regimen you were instructed to follow. The wagging finger isn't hard to visualize. The disease concept is also a useful tool for the insurance industry, because it defines and delimits the kind of treatment that will and won't be covered, for how long, and at what cost. Closer to home, most addicts' families (76 percent in a recent Gallup poll)[14] also see addiction as a disease, because it makes the disgraceful behaviour of their loved ones comprehensible and even forgivable. So the disease model becomes a convincing framework for understanding addiction from the outside—even when that definition is ineffective, inaccurate, or harmful for addicts themselves.

We should notice that the chief fortification for the disease model comes from statistics that are often misinterpreted or even deliberately distorted. David Sack, a psychiatrist and president/CEO of

Elements Behavioral Health, a chain of recovery centres, had this to say in a 2014 online debate in the *New York Times*:

> A recent study of heroin addicts found that at the end of one year, approximately 50 percent remained in treatment and more than 80 percent had used heroin regardless of the type of treatment they received. By the end of three years, only 8 percent were continuously abstinent. . . . These outcomes are not dissimilar to those observed in Type II diabetes, hypertension or asthma, where only a minority of patients achieve clinical control after extensive treatment.[15]

Yet a quick look at the source article from which he derives his statistics paints a very different picture. The "8 percent [who] were continuously abstinent" is a gross distortion. Here's a quote from the abstract of the paper where Sack got his figures: "The proportion who reported abstinence over the preceding 12 months, however, increased significantly from 14% at 12 months to 40% at 36 months."[16] In other words, three years after first being interviewed by researchers, 40 percent of heroin addicts had managed to quit using, without a single relapse, for a whole year. And heroin is perhaps the most addictive drug we know of. Why didn't Sack include this finding in his synopsis? Doctors aren't the only ones who misuse statistics, but we assume they will be more conscientious than most.

∾

I'm not the first or the last to argue against the disease definition. Recently, authors Sally Satel and Maia Szalavitz, both experts on addiction, have eloquently refuted the idea that brain change equals brain disease. Ivan Oransky, executive editor of Reuters Health, speaks out against the "mania for medicalization," as did Stanton Peele a generation ago in his 1989 book *Diseasing of America*. Oransky claims that those bucolic treatment centres, which I've noticed have names like Clearview, Clarity Way, and Promises, are "selling you on the fact that you need to be treated."[17] Current rehab experts, like the controversial William White, have poked large and small holes in the scientific foundation of the disease model, often citing the counterintuitive finding that most addictions end "spontaneously"—that is, without treatment, as noted previously. Harvard researcher Gene Heyman reframed addiction as a "disorder of choice" in his 2009 book *Addiction: A Disorder of Choice.*[18] More recently Heyman traced a "natural" (i.e., developmental) time frame of recovery for each of four addictive drugs: pot, alcohol, cocaine, and tobacco.[19] It's hard to square that kind of schedule with the notion of a disease that requires treatment. According to some experts, the best evidence against the disease model comes from the study of heroin-addicted veterans of the Vietnam War, about 75 percent of whom kicked the habit once they returned home. A number of us view this heartening statistic as the human counterpart to what Bruce Alexander demonstrated in his classic "Rat Park" studies.[20] Alexander and colleagues offered rats a choice between morphine solution and water. Rats raised in isolated steel cages chose the morphine. But once placed in a large wooden enclosure with other rats and allowed to socialize, they switched to plain water, even when they were currently addicted. In other words, they "quit" voluntarily.

The argument that addictive behaviour is actually a choice got another boost recently from Carl Hart's work at Columbia University. Hart, a dashing, dreadlocked associate professor, gives crack addicts the opportunity to choose between drugs and money once checked into a residential setting. In a controlled experiment, he offers addicts either a hit of crack or an IOU for money they'll receive a week or two later. Contrary to conventional expectations, they often take the money (as little as $5) instead of the pipe. Hart goes on to argue that the dispossessed young men of the Miami ghetto where he grew up had few attractive alternatives, which is why they chose drugs and why drug use is endemic to the inner city.[21] A similar message comes from the gripping 2012 documentary *The House I Live In,* which exposes the War on Drugs as a culture war whose victims are primarily poor, black Americans.

Yet I don't buy the dichotomy—that addiction must be a deliberate choice if it's not a disease. British journalist Peter Hitchens, brother of the late Christopher Hitchens, is one vitriolic commentator who, like several others who subscribe to the choice model, has little sympathy for the plight of addicts. In a recent televised debate on the BBC's *Newsnight,* he had this to say: "People have problems with drugs and drink. People like taking them and don't want to stop. It doesn't mean they have a disease." (See also the unambiguously titled *Addiction Is a Choice* by Jeffrey Schaler, who seems to bear a simmering contempt for addicts because they choose their own troubles.) The debate may hinge on the issue of control. If you don't have control over your substance use, then you have a disease, and if you do have control (but aren't using it), then addiction is a choice. The leak in the logic is the assumption that choice is a deliberate, rational function we can apply at will. But choice is nearly always irrational—which is only to say that it is executed by the same brain that gives rise to hope, need, fear, and uncertainty, a brain that's highly sensitive to learned associations and contextual cues, a brain that forges new connections based on the activation of existing connections and the strong emotions they render. So when I argue against the disease model, I'm not arguing that addictive behaviours are fuelled by voluntary choice. In fact, I don't see how anyone who has talked with deeply addicted people or read their agonizing memoirs can imagine that their behaviour is guided by simple volition.

And yet . . . and yet. A friend, colleague, and longtime recovered addict recently sent me this anecdote:

I made contact with an old mate the other day. I genuinely thought he was dead, having not seen him since 1998. We used to use together, same stuff in the same way involving the same chaos. Much the same as myself, he used for over 30 years, mainly injecting heroin and cocaine at a rate of over £1,000 weekly. Six years ago he decided he had had enough and just stopped. He had never been on a script, never been in detox or rehab and never been anywhere near a treatment service. All on his own he has been totally abstinent for over six years, got a little fruit and vegetable store and fixes motorcycles.

Finally, one of the fastest-growing fissures in the armour of the disease model is the recognition that behavioural addictions assume the same characteristics, the same trajectory, and often the same outcomes as substance addictions. Gambling, sex addiction, porn preoccupations, eating disorders, and even excessive Internet use have entered the spotlight next to drugs and booze: they too turn out to have serious consequences, including broken relationships, broken health, and sometimes death. What's most informative is that the neural consequences of behavioural addictions indicate the same cellular mechanisms and the same biological alterations that underlie drug addiction. The present trend is to classify these addictions as disorders, a vague label that overlaps considerably with "disease." But then where do we stop? Should Net surfing, hoarding, compulsive shopping, and unrequited love also be classed as diseases or disorders? It seems that the slippery slope gets increasingly slippery with closer inspection.

A FRESH LOOK AT THE BRAIN

Yet among the opponents of the disease model, no one has fought fire with fire and tackled its neuroscientific foundations. Like the general public, most of those arguing against the disease model assume that "brain change" automatically implies a disease process; then they change the subject. Others tune out (or get mad) when the brain is even mentioned in regard to addiction, because they assume that a neuroscientific description will somehow replace a more psychological or humanistic perspective, rather than complement it. It's as if students of addiction have to choose: either admit that the brain is a really important organ, in which case addiction is a brain disease, or put the brain back in the closet, in which case you can go on talking about choice, environmental factors, social anthropology, and all the rest of it. In his most recent book, Stanton Peele, a longtime opponent of the disease model, gives his readers a stark choice: either accept addiction as a brain disease, in which case addicts are powerless to fight it, or recognize that addiction is a personal, self-defeating habit.[22] Peele wants to reject the fatalism inherent both in the disease model and in some AA rhetoric. With this I concur. But we don't have to reject neuroscience or AA to do that. In fact, a fresh look at the brain can help.

First, though, we have to remove brain science from the arena of medical politics and connect it back to its natural partners, psychology and personal experience. That won't be easy. In a 2013 journal article, Satel and Lilienfeld echo Peele's challenge: "The brain disease model implies erroneously that the brain is necessarily the most important and useful level of analysis for understanding and treating addiction."[23] These authors are right to encourage multiple levels of analysis in our attempts to understand addiction. But must we step away from the brain in order to accept addiction as a complex human problem? Must we throw out the brain with the bathwater? As a neuroscientist, I find this impossible. I see the brain as fundamental to our humanity—and as fundamental to addiction. Whether we construe addiction as a disease, a choice, a complex sociocultural process, self-medication, or a string of bad-hair days, we only have one brain, and it's central to everything we do, everything we are. So a very important question is simply this: what does the brain *do* in addiction?

But before trying to answer that question, we need to understand how brains change normally. In fact, brains are supposed to change. Brain change—or neuroplasticity—is the fundamental mechanism by which infants grow into toddlers, who grow into children, who grow into adults, who continue to grow. Brain change underlies the transformations in thinking and feeling that characterize early adolescence. In fact, developmental neuroscientists estimate that "as many as 30,000 synapses may be lost *per second* over the entire cortex during the pubertal/adolescent period."[24] Brain change is necessary for language acquisition and impulse control in early childhood, and for learning to drive a car, play a musical instrument, or appreciate opera later in life. Brain change underlies religious conversion, becoming a parent, and, not surprisingly, falling in love. Brains have to change for learning to take place. Without physical changes in brain matter, learning is impossible. Synapses appear and self-perpetuate or weaken and disappear in everyday learning. Learning alters the communication patterns between brain regions and builds unique configurations of synapses (synaptic networks) that house knowledge, skill, and memory itself. The connection between learning and brain change has been studied for more than a hundred years: it was reasonably well understood by the 1940s, and the search for specific cellular mechanisms continues today. Whether repairing the damage caused by a minor stroke or altering emotional processes in the wake of trauma, neuroplasticity is at the top of the brain's resumé.

To repeat: proponents of the disease model argue that addiction changes the brain. And they're right. It does. But the brain changes anyway, at every level: gene expression, cell density, the concentration and location of synapses and their fibres, even the size and shape of the cortex itself. Of course, neuroscientists who subscribe to the disease model must know that brains change with learning and development. So they must view the brain change that accompanies addiction as extreme or pathological. In fact, they would have to show exactly that in order to be convincing. They would have to show that the *kind* (or extent or location) of brain change characteristic of addiction is nothing like what we see in normal learning and development, or even in the more extreme transitions people go through when they fall in love or have children. But that's where they step onto thin ice. The kind of brain changes seen in addiction also show up when people become absorbed in a sport, join a political movement, or become obsessed with their sweetheart or their kids. The brain contains only a few major traffic routes for goal seeking. Like the main streets of a busy city, the same routes get dug up and paved over time and time again, no matter who's in charge.

Brain disease may be a useful metaphor for how addiction *seems,* but it's not a sensible explanation for how addiction *works.*

TWO

A Brain Designed for Addiction

What's it like to be a brain? It's nothing like being a person. We carry our brains around with the rest of our bodies because we'd be lost without them, but brains are only parts of people. And we have many other parts with many diverse functions. So when we say my brain made me do it, or my brain hurts, or my brain really likes alcohol, we are mixing metaphors, confusing levels of analysis. Similarly, to state that craving or pleasure takes place in a certain part of the brain is a figure of speech. The so-called reward centre is actually a conglomerate of remarkably distinct cell assemblies. The fact that these cell assemblies become activated by the scent of an upcoming binge does not mean that a reward bell chimes pleasantly in that location. Brains don't have likes or dislikes, rewards or punishments, goals or cravings. Those are things people have. And brain cells don't contain thoughts or feelings. What they contain are membranes, molecules, proteins, blood, and constantly fluctuating levels of electricity. *We* crave, *we* have feelings, and *we* get addicted.

Brains just do what hundreds of millions of years of evolution have determined to be useful, and that includes identifying things that taste good or feel good *to us*. The brain distinguishes those things from everything else—the background music of the humdrum world—and propels us to go after them. Psychologists call those things "rewards," things like ripe peaches, fresh bread, orgasms, and even cuddles, and, by association, money, good looks, and power. Oh, and drugs. Addiction may be the uncanny result of a brain doing exactly what it's supposed to do.

Brains are nothing like digital computers either. We often compare the brain to a digital computer, because it's pretty good at solving logical problems. But the computer model of the brain died a natural death in cognitive science at least twenty years ago. From the early 1990s onward, cognitive scientists and neuroscientists began to describe the brain as "embodied" and to see its functions in biological terms. Around the same time, computer scientists began to put away their "artificial intelligence" machines and replace them with "neural networks"—which work completely differently. These models at least approximate the network-like quality of real brains, whose cells are connected more like ivy leaves than like boxes and arrows. And, all the better for realism, these models make mistakes!

I remember when I first learned how easily neural networks make mistakes. Three descriptor terms were fed into the network model as "input." Our task, as grad students, was to come up with a quick answer to the question "Who are we talking about?" and then compare that answer to the "output" provided by the network. The three terms were "movie star," "politician," and "intelligent" and the answer that came to everyone's lips was "Ronald Reagan." The network spat out the same answer. "But you see," our typically left-leaning professor said with a grin, "one-third of the information is false!" It pays to remember that brains make decisions based on biased, convoluted, and often just-plain-mistaken input.

Daniel Kahneman's bestselling book *Thinking, Fast and Slow* summarizes thirty years of progress through which psychologists (and students of behavioural economics) have come to recognize how biased and irrational our thinking can be. None of this should be very surprising. After all, the brain is a body part whose concern is the slaking of desires and avoidance of risk, goals carried out by hand, tongue, teeth, feet, and genitals. Rationality is a useful tool for planning our route in rush-hour traffic, for dinner-table debates, and for getting A's in school. But it's not the front-runner when it comes to eating, sex, pain, pain relief, self-fulfillment, looking cool, and US elections.

So to say that addiction isn't rational is just stating the obvious. The irrationality (including self-destructiveness) of addiction does not indicate that the brain is malfunctioning, as it would if diseased. It just shows that it's a human brain. Thinkers from Homer to Dennett and writers from Shakespeare to Nabokov have made it abundantly clear that irrationality is an essential feature of being human.

PLASTICITY AND PERMANENCE IN BRAIN CHANGE

Human brains and reptile brains have a lot of the same parts. Those parts are more sophisticated in human brains, but they still retain most of the same functions. Subcortical regions like the hypothalamus and brain stem get us to act rapidly—to fight, grab, flee, eat—in direct response to input from our senses and without much thought. What distinguishes mammalian brains, and especially human brains, is that they live in environments that are far more complicated than the stomping grounds of lizards. We mammals are quick to adapt to those complications. We don't just die when it gets too cold outside. We burrow underground or turn up the heat. And when a certain type of food is no longer available, we whip up a sandwich or stop at McDonald's rather than starve to death. The reason mammals have this dexterity and adaptability is that their brains are designed for learning—they are designed to change—in synch with their environments. Lizards don't learn very much. Their repertoire of skills is innate. But humans learn almost everything they ever come to know or do. That's why babies are completely helpless. They can't do a thing, because they haven't learned how.

So reptile brains are almost entirely prefabricated. But human brains require extensive cellular changes, from before birth to the end of our lives, in order to function at all. Most of these changes take place in two broadly defined areas. The first is the cerebral cortex, that vast surface of braided greyish brown matter that covers the innards of the brain with networks of "programmable" cells. The second is the so-called limbic system, which includes the amygdala, hippocampus, and striatum—regions that play a major role in emotion, memory, and goal pursuit. Cells in these regions are also programmable for the most part. The more than twenty billion neurons in the cortex and limbic system aren't told which other cells to connect with when we're in the womb. Although the initial placement of neurons is similar for all human brains, the connections among them—the synapses, which number in the trillions—are designed to change radically. They do this throughout life, in response to our experiences. And each wave of synaptic change alters the *way* we experience things.

That's a neat little trick: the way we experience things shapes our biological matter, and those biological changes shape the way we experience things subsequently. In other words, changes in brain structure make that way of experiencing things more available, more probable, on future occasions. This can take the form of a self-reinforcing perception, an expectancy, a budding interpretation, a recurring wish, a familiar emotional reaction, a consolidating belief, or a conscious memory. They're all different forms of "permanence"—of the way brain patterns settle into place, so that traces of the past can shape the present. What I'm describing is a feedback loop: between a way of seeing, remembering, or acting on the world and a structural change that perpetuates that way of seeing, remembering, or acting. Thus the mind and the brain shape each other. And ordinary classroom learning is just one version of this more general phenomenon—a brain that changes itself (a phrase captured by Norman Doidge in the title of his 2007 book).

There's an important addendum to this big picture, and it's fundamental for understanding addiction. When our experience of the world is fraught with strong feelings—whether of attraction, threat, pleasure, or relief—brain change takes on extra momentum. What drives this momentum? Emotions focus our attention and our thinking, and particular emotions (in response to something) call up particular thoughts and behaviours, thereby fuelling the same feedback cycle every time that something is encountered. When those emotions recur over and over, with each repetition of the feedback cycle, our overly focused brains inevitably change in a particular direction, entrenching a certain emotional experience a little more each time. Most relevant to addiction, the feeling of *desire* for something specific shapes the brain more acutely than other feelings. As you will see, desire-laced experiences mould the brain into a vehicle for creating similar experiences, also rooted in desire, for a long time to come.

The brain would be useless if it weren't highly changeable, highly sensitive to events in the world. But since we need stability in our percepts, concepts, and actions, in order to get through the day and make provisions for tomorrow, brain changes almost always settle into habits. And once formed, habits—even minor habits—remain in place, sometimes for the rest of our lives. Examples range from idiosyncratic patterns like nail biting and suspiciousness to cultural norms like politeness and sexual stereotyping. New neuronal pathways, and corresponding patterns of thought and behaviour, start off tentative and fluctuating. But after they've been activated repeatedly, fledgling pathways get more entrenched, more con-cretized, and eventually carved in stone, or at least in flesh. Thus brain changes naturally tend to stabilize and crystallize. And if new changes come about, they tend to *re*stabilize. It wouldn't do to have a brain that fluctuated unpredictably with every passing event. So change and stabilization go together. In a word, that's learning. It's also a second critical point when it comes to understanding addiction.

William James, one of the first and most brilliant modern psy-chologists, had this to say about "habit loops" over a century ago:

> What is so clearly true of the nervous apparatus of animal life can scarcely be otherwise than true of . . . the automatic activity of the mind. . . . Any sequence of mental action which has been frequently repeated tends to perpetuate itself; so that we find ourselves auto-matically prompted to think, feel, or do what we have been before accustomed to think, feel, or do, under like circumstances, without any consciously formed purpose, or anticipation of results.[1]

WARNING: MAY BE HABIT-FORMING

The word "neuroplasticity" is on everyone's lips these days. The term simply describes brain changeability and elevates it to a first principle. Which makes sense: there's nothing more fundamental to the human brain than changeability. Yet neuroscientists who study addiction seem to have missed the point. They put people through a number of brain scans, and when they notice changes after someone has taken a lot of cocaine or drunk a lot of booze, they say, "Look! The brain has changed!" If neuroplasticity is the rule, not the exception, then they're actually not saying much at all. The brain is supposed to change with new experiences. In fact, the newer, more attractive, and more engaging something is, the more likely the brain is to change, and the more likely those changes are to condense into habits—an outcome of more frequent repetitions.

People have referred to addiction as a habit throughout recent history. That's just what it is. It's a nasty, often relentless habit. A serious habit. An expensive habit. But what makes it so enduring, so relentless, so difficult to change? What makes it different from what we might call more benign habits? Three things. First, it's a habit of thinking and feeling—a mental habit—not just a behavioural habit. It's easier to stop singing in the shower than it is to stop seeing the world as violent or unfair. Second, the feeling part of addiction always includes the feeling of desire, which is of course the theme of this book. And third, it's a habit that becomes compulsive—a topic that will be explored in detail later. Perhaps all habits, once formed, are compulsive to some degree. The brain is certainly built to make any action, repeated enough times, into a compulsion. But the emotional heart of addiction—in a word, *desire*—makes compulsion inevitable, because unslaked desire is the springboard to repetition, and repetition is the key to compulsion.

Like all habits, addiction quite simply grows and stabilizes, in brain tissue that is designed (by evolution) to change and stabilize. Yet addiction belongs to a subset of habits: those that are most difficult to extinguish. To understand addiction, we need to see it as the outcome of a normally functioning brain, not a diseased brain. Still, we must acknowledge that it's an extreme outcome, and that's what has to be explained.

Not all habits start with desire or attraction. Anxiety and other negative emotions can cultivate new habits too. Nail biting is one example, but so are gaze aversion, classic defence mechanisms such as rationalization and perfectionism, and the habitual avoidance of certain people, places, sexes, or races. Personality patterns based on anxiety arise from a slightly more complicated feedback loop, involving anxiety (or shame, or outright fear) and escape, which is a kind of reward—or else an ongoing state of vigilance, which isn't rewarding at all. Avoidance of possible threats isn't fun, but the emotion of anxiety can strengthen habits of avoidance until they overtake the rest of the personality. And maybe that beats overwhelming anxiety. As you will see, the habits of desire that characterize addiction are often intermingled with habits born of anxiety or shame.

IT ALL COMES DOWN TO FEEDBACK

Exactly how *do* brains develop? And how do they develop habits? Development is not a simple concept. Plants grow, and so do brains, ecosystems, corporations, and climate patterns. But plants grow along the lines their genes have dictated. Cells divide and follow instructions, so oak trees and tulips look very much like their parents. In contrast, human brains, ecosystems, corporations, and climate patterns grow unpredictably, filling the world with enormous diversity. They create their own fate rather than fall into a fate that's already set out. They grow, not just by following pre-specified guidelines, but by a process known as *self-organization*. They organize themselves, changing their own structure as they go. As described earlier in this chapter, the patterns we find in brains (and communities, and the weather) are *self-perpetuating*, building on themselves over time. They change course without much notice (though we might argue that climate change is giving us plenty of notice), and then they continue to grow in *that* direction. And then they stop changing, or at least they settle and change a lot less. Like the ruts and rivulets formed by rainwater in the garden, they start off unpredictable, but they create and consolidate their own "fate" as they grow. They stabilize. They form habits.

What causes this kind of growth? The answer, surprisingly, lies in the bountiful heart of the feedback loop itself. Feedback loops are just about the most powerful thing in the world. For example, feedback between increasing energy production and increasing energy demand is what causes climate change. Cancer is a nasty feedback loop between increasing cell growth and decreasing inter-cell signalling. When Jimi Hendrix got his guitar pickup (an input system) close to his amp (an output system) he set up a feedback loop, enabling him to create sounds guitars don't usually make. In *my* attempts to reach my listeners, auditory feedback has generally been a nuisance. Instead of a new art form, I get a loud hum and a roomful of irritated students. Feedback simply makes things grow. It doesn't care what.

Back to the brain. The feedback loop I mentioned, between the experience of desire (or other strong emotions) and ongoing changes in brain structure, is one of the prime engines of neural self-organization, or human development in general. If you bite into a piece of chocolate cheesecake and feel a gush of pleasure, and if you're left with a lingering desire (which surfaces the next time you eat out), you have begun a trajectory of "loving chocolate cheesecake." At first, maybe just a few synapses have been altered by such an experience. But those changes increase the chance that you will encounter chocolate cheesecake on a future occasion. You will notice it more, acquire it more, and eat it more. And before long a string of subjective experiences will give rise to a string of neural changes that continue to feed back to the ever more predictable consumption of chocolate cheesecake. Congratulations: you have now developed a habit.

It's crucial to remember that it's not just attraction or desire that fuels feedback loops and promotes neural habits. Depression and anxiety also develop through feedback. The more you think negative thoughts or scary thoughts, the more synapses get strung together to generate scenarios involving loneliness or danger. *Who else might not like you? Did you really think you were going to get away with that?* Or generate strategies—often unconsciously—for dealing with those scenarios. *I'll have to be extra nice if I don't want to be rejected.* The result is the sculpting of neural flesh into a breeding ground of personal habits. And one way to capture the combination of those habits is with the word "personality." Which means I could summarize who you are, in the simplest terms, by listing your most distinct habits, especially those evoked by negative emotions. Isn't that what we do when someone asks us, "What's he or she like?"

Neural patterns forged by desire can certainly complement and merge with those born of depression or anxiety. In fact, that's an important bridge to the self-medication model of addiction. In his 2010 book, *In the Realm of Hungry Ghosts,* Gabor Maté persuasively shows how early emotional disturbances steer us toward the relief that becomes addiction. So when we examine the correlation between addiction and depression (or anxiety), we should recognize that addiction is often a partner or even an extension of a developmental pattern already set in motion, not simply a newcomer who happened to show up one day.

Different experiential feedback loops create different parts of us, based on different brain patterns, like the work of a painter who does one section of a painting, then another, then another, then goes back and works some more on the first. Of course it wouldn't be a very good painting if those parts didn't hold together with some degree of coherence, and the painter knows this very well. When it comes to human development, we don't necessarily believe that there's a painter in charge. But there is still a lot of coherence in the end result, and that coherence is often what we call personality. Being this sort of person or that sort of person simply requires a few different feedback loops evolving together, in leapfrog jumps or in continuous collusion. Some of these feedback loops are powered by attraction (like the tendency to be generous because you're in love) and some by anxiety (like the submissiveness motivated by fear of rejection). These feedback loops work together. They support each other, like a couple of drunks.

So repeated experiences establish patterns, forming habits, and those habits link with other habits that also evolve with repeated experiences. And though developmental outcomes are unpredictable at first, we *can* predict that whatever, or whoever, emerges will endure and solidify as time goes by—a finished (at least partly finished) human product. Much like a painting. When you gaze upon your one-year-old child, you won't be able to guess what kind of person he or she will become. But you can rest assured that this person will have a distinct personality. When you gaze upon your thirteen-year-old entering high school, you won't be able to guess what kind of young adult will come out four years later. But you can bet that he or she will be pretty much a certain kind of person (even if that includes being mixed up) and the same kind of person from one day to the next. Let's hope it's a kind of person you like.

Now here's the point when it comes to addiction. If everything I just described is more or less correct, if this is how humans develop, how they *form*, then you don't need an external cause like "disease" to explain the growth of bad habits, or even a set of interlocking bad habits (like being a drug addict and a criminal and a liar). Bad habits self-organize like any other habits. Bad habits like addiction grow more deeply and often more quickly than other bad habits, because they result from feedback fuelled by intense desire, and because they crowd out the availability or appeal of alternative pursuits. But they are still, fundamentally, habits—habits of thinking, feeling, and acting. The brain continues to shape itself with each repeat of the addictive experience, until the addictive habit converges with other habits lodged within one's personality. Take, for example, the habit of anticipating trouble, feeling anxious, and searching for relief. New habits may spring up around it: you may find you don't have a lot of friends left, and you're not on good terms with your mom and dad, because you've become consistently snarly or distant. And if you ended up lying to your girlfriend about what you were doing all evening, that will probably develop into a habit as well.

One reason the end result is so coherent is that the social habits that accompany addiction, like lying to yourself and others, mesh so very easily with the self-soothing (or self-feeding) habits that make up the addiction itself. Habits fuelled by anxiety and shame continue to evolve with habits fuelled by desire, because addiction is a risky business. Addicts take big chances, hide the distasteful things they do, and hope for the best. In that respect, addiction can be seen as an extension of personality development. And it can surely be seen as self-medication, but the anxiety and shame that have to be medicated result more and more from the addiction itself.

NETWORKING NEURONS

The main thing brain cells do is send and receive electrochemical energy, and they do so through their connections (their synapses). Each time a neuron fires, which could be once a second or up to a thousand times a second, a spurt of electrochemical energy is passed along to the neurons it connects to. This energy flow is how experience is produced by the brain. That's what puts the picture on the screen of your mind. But of course the brain needs input from the outside world in order to experience *something*. As highlighted by fifties-era sci-fi, a brain in a tank is not a happy brain.

Synapses are tiny spaces between neurons, where the fibres of one cell (the sender) connect to the fibres of another cell (the receiver). These way stations are where molecules cross over from the sender cell to the receiver cell, and when enough molecules get across (usually resulting from a group effort by many sending cells), the electrical charge of the receiving cell changes. This change in electrical charge results in an increase (called excitation) or a decrease (called inhibition) of that cell's firing rate. And that change in firing rate can influence the next neuron in line, making it more or less ready to send its energy to the following neuron, and so on. The result? A *synaptic pathway:* a train of neurons, connected by synapses, each affecting the next in line.

That's how neurons communicate, but how does all this activity result in long-term change? Changes occur almost entirely at the synapses—including changes in the shape of current synapses, the sprouting of new synapses, and the decay and eventual disappearance of old synapses. These changes affect the *degree of connection* between neurons. Neurons either become more connected (so that more molecules get through) or less connected (so that fewer molecules get through). And these structural changes actually result from the electrochemical flow that produces our moment-to-moment experience. But these structural changes are not guaranteed. Sometimes they happen—laying down new memories, new learning—and sometimes they don't. As my mother used to tell me, I often don't learn from experience. How could that be?

The emotional intensity of the experience, the degree of focus and concentration, and the recurrence of that experience over time are usually what make the difference between simple experience and *learning.* If your boss's recent remark sent a flock of molecules from neuron X to neuron Y, enough to change its firing rate, then you are likely to perceive something: *there he goes again.* But if that remark was caustic, or upsetting, or heard once too often, then the connection between those two neurons is likely to be strengthened, so even more molecules will cross between them the next time around—the next time your boss says something even slightly unkind. That's brain change. (Real brain change always involves groups of neurons; I use "neuron X" and "neuron Y" to simplify the description.) Brain change equals synaptic modification, and synaptic modification results from synaptic activity that is boosted by emotion, attention, and repetition.

As Donald Hebb said in the 1940s, *what fires together wires together.* Cells that fire each other end up being more strongly connected—you could even say their connections become "hardwired." That's how learning takes place, but *emotional significance* is the switch that gets it started and keeps it going. If it doesn't mean much, if it doesn't induce feelings, it's not going to capture your attention and it's not going to get recorded in synaptic structure. Just ask the kid who's been looking out the window instead of paying attention in class. And *repetition* is the engine that strengthens and perpetuates learning, changing it from a temporary gloss to a deep engraving of the world. Which is why the teacher not only raises her voice (to compete with whatever's outside the window) but also says the same thing eight times in a row, and several more times next week.

So how and when does learning take place? Learning usually requires at least moderately strong emotions, and it is greatly facilitated by repeated experiences (or else one or two powerful experiences). These create the first layer of a new synaptic pattern, a new lattice of synaptic connections. Novel patterns show up as networks—villages or towns connected by dozens of little roads. But the networks become more robust and more efficient with repetition, and the learning gets deeper. Think of the dozens of little roads being replaced by several main roads and maybe eventually a freeway.

This formula for learning is epitomized by addiction. Your first snort of cocaine probably produced a novel firing pattern. (If not, you'd have tried a second snort or found another dealer.) Then, each time you (not you personally, of course) snorted coke, more synapses were changed, reinforcing this firing pattern, this "cocaine" configuration. That configuration would soon connect regions all over your brain. These include parts of the cortex—the perceptual cortex, in charge of seeing and hearing; the prefrontal cortex, in charge of thinking and planning; and the motor regions, in charge of putting those plans into action. But they also include the limbic regions involved in feelings and motives—the amygdala and hippocampus, as well as the striatum (which is not usually defined as "limbic" per se but . . . close enough). So it's more or less the whole brain—the parts involved in thought and perception and the parts dedicated to feelings and instincts—that gets included in the "cocaine" network. Which is why thoughts, feelings, and action patterns change and crystallize together. To repeat: it's the whole brain that programs itself, self-organizes, develops a habit—a habit that eventually becomes integrated with all your other habits.

New habits—new networks—are like cliques that emerge from the interactions of multiple users, such as Facebook participants. Busy synapses connect certain neurons to other neurons, forming and strengthening some networks, while other networks shut down with disuse. So an ensemble of neurons, connecting different brain parts, forms a single learning trajectory—a "cocaine" configuration, or alternatively a configuration for being an amazing dancer, skier, or chef. Yet when new synaptic patterns emerge, older patterns don't just disappear. They get modified. They may fade with disuse. But they may also become components of the new pattern, or act as conduits to the new pattern, like road signs pointing toward a newly constructed overpass. For example, ruminations about getting high can lead to strategies for self-distraction—components of a network for successful recovery. But they can also lead to thoughts of leaving home. The big picture is that habits at one point in development influence habits at the next point, which influence habits at the next point, and so on. Which means something bad and something good when it comes to recovery. The bad thing is that there's no way to wipe the slate entirely clean. The good thing is that recovery can build on lessons learned in addiction, and it can build on itself over time.

IT'S NOT UNUSUAL

To say that addiction changes the brain is really just saying that some powerful experience, probably occurring over and over, forges new synaptic configurations that settle into habits. And these new synaptic configurations arise from the pattern of cell firing on each occasion. In other words, repeated (motivating) experiences produce brain changes that start to define future experiences—at least those in the same realm. So getting drunk a lot will sculpt the synapses that determine future drinking patterns. Whether it's sacramental wine or beer at the ball game, it'll soon become part of the same familiar drama, if it isn't already. (It probably won't affect how you feel about your grandmother or play with your dog—though it might.) These changes don't *result* from addictive substances. They are not *caused* by booze or drugs. They result from having a string of similar experiences. Nice experiences. Experiences of relief. Experiences that feel good, or at least better than the rest of your boring and depressing life. These brain changes are caused by motivated repetition—repetition of something special—and how the brain responds to it. The powerful experiences that get the ball rolling are simply events that affect us deeply. Because they are engaging. Because they mean something. As they become even more meaningful, the corresponding brain changes gather more momentum, building on themselves, digging their own ruts—rainwater in the garden.

Experiences that change the brain most rapidly or extensively *might* involve drugs or alcohol, and it's no accident that those substances can be highly addicting. Alcohol and heroin would certainly be less addictive (and a lot cheaper) if they led to experiences that are boring. But high-powered brain-changing experiences also include gambling, binge eating, having a lot of sex, sitting back and watching other people have a lot of sex . . . or simply falling in love. There is nothing more stimulating, salient, attractive, and compelling than the face (and body) of your first high school crush. How many times a day did that person register on your consciousness?

~

That's enough neuroscience for now. It's enough to understand the basic features of brain change, how brain change leads to the formation of habits, how recurrent experiences continue to sculpt synaptic networks that continue to pull for similar experiences, and how motivation boosts each step of this developmental sequence. In the case of bad habits, habits that are most difficult to break, alterations in brain flesh have nothing to do with rationality, though they may be guided or misguided by the things you say to yourself in your most rational voice. And they happen regardless of the specific content of the experience, the flavour of the reward you're after (e.g., sex, drugs, rock and roll). They take place through changes in the way brain cells connect to each other, by the strengthening of some synapses and the weakening of others. They take place in many brain areas, which means that single habits can be quite complex, integrating thoughts, feelings, and behaviour patterns. And addictive habits tend to couple with other habits, simply because they work well together.

The last thing to say is that the same basic mechanism of habit formation—change and stabilization of synaptic networks—works differently in different parts of the brain. For example, the emotional side of habit formation emerges from changes in the amygdala and its close ally the orbitofrontal cortex; new action patterns take root in the motor regions, including regions that organize actions and those that execute them; new habits of thinking are centred in changes in the upper regions of the prefrontal cortex; and new sources of attraction and desire—which are obviously central for addiction—derive from changes in the striatum and its neighbours. These different regions, and their specialized functions, will be highlighted one by one over the next five chapters. I'll talk about what these brain regions do and how they grow new patterns, to help make sense of the lives of the people you will meet—when they became addicted and when they began to recover. And I think you'll see that the brain changes that underlie addiction and recovery are more normal than abnormal, though their outcomes can be extreme. Addiction may be a frightful, devastating, and insidious process of change in our habits and our synaptic patterning. But that doesn't make it a disease.

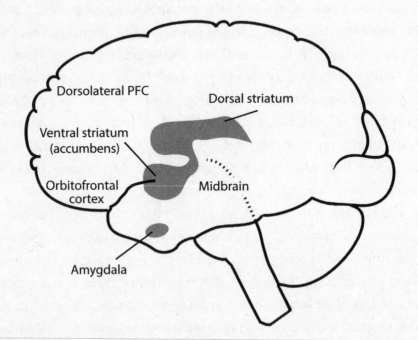

Figure 1. A sketch of the brain, showing regions most relevant to addiction.

Limbic and associated structures: the "motivational core"

Ventral striatum (accumbens): the "southernmost" part of the striatum, responsible for impulsive actions leading to goals, feelings of attraction, desire, anticipation, craving, and some aspects of reward itself; fuelled by dopamine sent from the midbrain.

Dorsal striatum: the "northern" part of the striatum, activated when goal-directed behaviours shift from impulsive to compulsive; central to stimulus-response learning; triggers actions that are automatic and difficult to turn off; also fuelled by dopamine.

Midbrain: contains cells that send dopamine to diverse parts of the limbic system and cortex, including the striatum, amygdala, and prefrontal cortex.

Amygdala (*pl.* amygdalae): a pair of small clusters, one on each side of the brain; acquires and maintains emotional associations, triggering the same emotion on subsequent occasions; focuses attention on the likely source of this emotion.

Orbitofrontal cortex (OFC): bottom surface of the prefrontal cortex; closely connected to the amygdala and accumbens; uses signals from these regions to create context-specific interpretations of highly motivating situations; generates expectancies and helps initiate an appropriate response.

Prefrontal cortex (PFC):
regions involved in self-comprehension and self-regulation

Medial prefrontal cortex: inside surfaces (on both sides) of the PFC; crucial for self-awareness, identity development, and interpreting others' thoughts and feelings.

Dorsolateral PFC: outer/higher-up region of the PFC (on both sides); matures gradually with cognitive development; responsible for bringing memories to mind, sorting and comparing them, and using insight, judgement, and logic to fashion new perspectives, make decisions, and adjust previous decisions; I call this "the bridge of the ship."

THREE

When Craving Comes to Power

Natalie's Story

Natalie grew up in a town in the eastern United States and, like a lot of other kids of her age and circumstances, ended up going to a small liberal arts college in a nearby town. She describes herself as a pretty competent person, smart enough to get by. She could pick up the gist of a situation without obsessing over it; she knew how to stay out of trouble. She knew how to relax in social situations—or at least look relaxed. And by the time she got to college, at age eighteen, her peers found her generous, easygoing, fun to be with. That's how she described her younger self during our first couple of interviews. And it fit my impression of her, even now, in her mid-twenties. It wasn't difficult for Natalie to make friends, and she could pick and choose the people she wanted in her life.

Natalie lived in a dorm for her first year of college, then moved to a shared apartment at the beginning of year two. She'd found a willing housemate in Grace, who liked being around Natalie, liked her quirkiness, her funny little hobbies like knitting and watercolours. She and Natalie would sometimes spend an afternoon painting together. But Grace faded into the background when Natalie found a boyfriend: a mild-mannered, sweet-looking guy named Fred. She'd met him at the restaurant where she worked three days a week, serving the kids she sat beside in Introductory Philosophy and Twentieth-Century Literature. She joked with him about how grandiose those students were, broadcasting their seasoned opinions about everything and everyone; then the two of them chuckled about their own grandiosity, putting everyone in their place. Fred was a pal, then suddenly a lover. That was new for Natalie, but it was also about time. Sex with Fred was friendly. And that was an achievement too—intimacy rather than subordination.

A couple of months after she and Fred got together, she brought home another friend, Steve—tall and lanky, a marijuana leaf tattooed on his forearm and a country-boy accent. There was something about Steve she found mysterious and attractive, but he was definitely not boyfriend material. Grace liked Fred well enough, but she wasn't particularly thrilled with Steve. He seemed to bring out Natalie's dark side, something vulnerable and spooky that didn't match the rest of her. Something that seemed to break through the surface more and more as the months went by.

But Natalie was still a good kid. Everyone thought so. Which is why nobody—not Grace, Fred, or even Steve—could have imagined that Natalie would spend nine months in a maximum-security prison before she got close to graduation.

~

Natalie had no particular moral compunctions about drugs, and she, Fred, and Steve experimented, like many of their friends, with whatever was going around. Magic mushrooms, even LSD once in a while—these were tickets to an exotic Disney World that you could visit without much planning or expense. The trip started and ended in a mere eight hours, ideal for strolling through the park and watching the sunlight devolve into rainbow highlights, so beautiful they sometimes made you catch your breath. And then there was Ecstasy, great for dancing at one of the two local rock cafes, where the DJ and the kids joined forces, coaxing great gouts of physical intimacy from the union of music and motion. These were "good" drugs. They didn't hurt you, nobody got addicted to them, they weren't very expensive, and you could still function the next day—though perhaps not with utmost clarity.

Heroin was different. I found it hard at first to imagine how this darkest of drugs had insinuated itself into Natalie's world. According to her recollections, the way was paved by a budding infatuation with prescription drugs, like the OxyContins Steve came up with. OxyContin, Percocet, Dilaudid—these are pharmaceutical opiates, designed to relieve pain. But Natalie and Fred found them to be just about the most pleasant high imaginable. They didn't pitch you into a colourful fairyland, the way mushrooms and acid did. Instead they wrapped you in a stocking of inner peace, utter relaxation. Not the kind of sedation you'd get from a tranquilizer, but something subtler and yet more potent. A feeling of well-being that didn't result from suppressing everything, making you drowsy and dopey. Opiates relaxed you by abolishing the sensation of threat and letting your mind wander freely in the fanciful landscape left in its place.

Natalie hadn't thought about it much, but a background hum of potential threat had been part of her world for years. It was hard to define, she told me, but it felt like a risk of *failing* one way or another—academically, maybe, or more likely socially. That was what mattered most, what had mattered since before high school, when she'd become a semiregular victim of teasing about being "dumpy," as she put it. Nothing terrible, but hard to tolerate nevertheless—an ongoing threat that got under her skin, a corrosive anxiety about being rejected by classmates who seemed like friends but might mutate into persecutors by next week. And even now, as a sophomore with a job, a boyfriend, and reasonable grades, she could imagine losing what she'd managed to gain. It could still happen. Some misty layer of anxiety was always floating just above the surface of things. Until opiates took it away.

So when Steve got a new batch of Oxys, which seemed to happen more and more often, Natalie felt a glow of anticipation, as if she were going home for the weekend—not her real home, but a dream home of soft edges and benevolent beings. That's when the synapses in her brain began a new wave of growth, forging a new set of connections. And when those synapses were activated, it felt to her like an approaching brightness, an arrow of hope. Fred liked the Oxys too, though she wasn't quite sure why. Natalie and Fred would purchase enough for a few days—a few days of that bright, peaceful high—and enjoy it together after classes or after work. But opiates affect the body, not just above the neck but also below it, stilling the stomach and slowing the breath and pulse. So it started to feel uncomfortable to go without them after a few weeks. Something went missing not only from her mood but from her body's familiar rhythms. We can call that a *physical addiction*, over and above Natalie's budding psychological addiction. (Though, of course, everything is physical. The brain is part of the body too.) The other problem was tolerance, another sign of so-called physical addiction. By the time spring came around, Natalie and Fred had to take three or four 20-milligram pills to feel high. And that got expensive, even when they crushed up the pills and snorted them. Nobody was making enough money to keep that going and still pay the rent on time.

Heroin was a lot cheaper, and Steve looked pretty cocky when he tossed a couple of $20 bags on the table one March afternoon. Fred and Natalie smoked it in a hastily renovated hash pipe, following Steve's example. But snorting it worked just as well, and that became their preferred MO in the following weeks. The high was actually quite a bit more intense than what they got from Oxys, but okay, bonus prize. An almost shocking river of peace would wash over her just a few moments after getting the heroin into her lungs or her mucous membranes. And there was something better than before to look forward to next time around. Synapses kept sprouting; neurons kept connecting. This was a fun ride, and the fact that nobody was steering made it even more exciting.

Still, heroin didn't dominate Natalie's life. "I had a lot of other things to think about," she told me, including her grades, her job, the balancing act she had to maintain with her parents on weekends and holidays—all the more difficult now that her stepdad's tantrums erupted so frequently. Heroin provided a welcome diversion, a reward at the end of a tough day, a psychic picnic table that she, Fred, Steve, and a few other friends could return to again and again. But she didn't *need* it. Not yet.

Then she tried shooting it. She watched Steve's girlfriend enact this grisly routine and was instantly intrigued. She pulled the woman aside and asked her to show her how to do it. And Natalie, in her usual thorough and responsible manner, memorized every detail of that intricate ritual. Pouring the right amount of brown powder into the spoon, adding water, holding the spoon steadily over the front burner of the stove until it bubbled for a few seconds. The little cotton ball that was necessary to filter the solution. The tourniquet. The bulging of the vein on the inside of her arm. It all had to be undertaken calmly and cleanly, nothing spilled, nothing spoiled. Until the needle penetrated the vein and she would feel, at that second, that there was an open passage for this exotic suitor, this ambassador from Afghanistan or wherever it was, to enter her body, her heart, her brain.

Then there was nothing more to worry about, from the moment the needle pierced the vein, even before the drug transformed the chemistry of her nervous system. That wash of peace started with the knowledge that this was a sure thing, no longer a maybe. This was what she'd been looking for, perhaps for a very long time.

∿

For Natalie, it wasn't drugs in general, or opiates, or even heroin that changed her life. It was the practice of shooting up. First came the feeling of being overwhelmed, as if by a breaking wave, a half-hour of chemical aftershock. That was cool. And so was the way the parts came together to make a whole, all those details converging into something unique. "That whole ritual thing . . . I loved it," she told me. "I would sometimes shoot up other people, because I was just better at it than others, and I enjoyed doing it for them." It seems Natalie was immersing herself in a dark liturgy. The skin and the steel, the bluish tint of the vein. The spoon and the flame, that moment of bubbling alchemy that marked the entrance into another dimension. And when the needle penetrated her own body, the feeling of the drug itself. The taste of it from inside. The click of shackles released all over her nervous system, reaching back to smooth every jagged memory, and forward, a guarantee of the hours to come, a guardian against harsh thoughts and images, the settling in of a consciousness infused with peace.

This harmony of sensations took up residence in Natalie's mind as networks of synapses converged in her brain. And from that headquarters a heightened urgency began to rule. It took less than a month from the time she started injecting heroin until her days became sloped in a singular direction, each afternoon a curled surface that would wrap into the shape of a funnel, a downward slanting tube that must empty into the reservoir of the moment. *That* moment—the moment of shooting up. Inevitable, from the first lurching rotation of her mind's eye to a point in time still hours away.

She'd be at the restaurant, working her shift, actually enjoying the summer sun streaming through the plate-glass windows, when suddenly it would strike. The thought, the image. It might be triggered by smoke rising from the cigarette of someone standing just outside the entrance. Or by the peculiar flattening of a spoon sitting next to the dirty plate on a nearby table, calling to mind the misshapen spoon of the dope-shooter, bent so that the bowl remained horizontal when the spoon lay at rest. Or by the vibration of her cell phone, acknowledging a call from Fred, who might already have been in touch with Steve. Who might already have enough for all of them. For tonight. Or by the gurgle of her stomach, which wasn't working very well these days. Or by the sight of her own arm. With her sleeves rolled down, no one could tell. (Though everyone seemed to know that Natalie was living on a different planet.) But when she rolled up her sleeves to wash, she'd often catch sight of the dark penumbra of a bruise centred at the seam inside her elbow.

Whatever the trigger, the ride each time was a replay of her previous rides, that sensation of being catapulted from wish to deed. First came the sudden tension in her stomach and chest, a lurching sensation of something imminent, a challenge or threat or golden opportunity, depending on how she played her hand. For a second or two the feeling had no colour, no content, and then it morphed into craving. She could not stop thinking about it from that moment onward, she recalls. It invaded her reverie, her every thought, while she went about the practiced motions of her work. She checked her phone every five minutes. She was sure that Steve would still be at home. It wasn't past one yet. She'd send him a text message as soon as she got to the kitchen. But what if he'd already scored? He might be on his way back now. In Natalie's words, "If he got stalled in traffic, it would be excruciating. . . . I would literally stare at the phone, willing it to ring." And when he did get back, would he have enough for all of them? What if he only had enough left for himself? Then she'd have to plead with him.

This kind of internal dialogue came out clearly in my interviews with Natalie. After some probing on my part, she remembered her thoughts as well as her deeds. She would rehearse the next phone call: C'mon, Steve. I'll go with you if you want. I don't mind. I'm off work at five. And she would imagine the bleakness of failure: He likes my company. But what if he's going away for the weekend? Shit, we should have saved enough. We put aside one bag, but that's not nearly enough for both of us. Well, fuck Fred in that case. He's always relying on me to take care of this. But what if he's already called Steve? What if he forgot that I'm going home Friday night? I need to get enough for the weekend; otherwise it's going to be totally awful to be with Mom and Shithead.

It was impossible to ignore the moment inscribed in the future of this day, the moment when the drug would be in her grasp. Until then, the pit in her stomach—anticipation mixed with desire mixed with dreadful anxiety—would not lessen. It would not soften. Not until she knew for sure. And even then, something could always go wrong.

By May, Natalie was not yet so physically addicted that she couldn't get through a day without heroin. But she was on her way. Here's how she describes her worries: "I knew I might be mildly uncomfortable tomorrow morning, and sure, it would get worse by evening. Two days without—hard to know." She hadn't gone two days without for weeks now. "It would probably feel like shit." And yet she didn't think about her addiction as physical. In her words: "The physical addiction was never foremost in my mind. It was far more the psychological. I was getting obsessed with it. If we didn't get it on a particular day, we would be up really early or up late that night, waiting for the phone to ring." In her fantasies, she composed a peculiar drama: I'm a drug addict, that's who I am. My life is a movie, and I'm the raunched-out chick playing the blues in the bus station, hunched over, tragic. Not the quiet little Natalie who reads a lot of books and doesn't get out much. She was living on the edge, no question, and that gave her life contour, thrust, colour, meaning—and this precipitous anxiety, this unquenchable craving, this quiet desperation that got louder by the hour as she waited for Steve to get back to her. What if his phone's out of juice? Could he have forgotten it at home?

∽

There are parts of the brain wrapped deeply within the cortical layers that we share with animals going all the way back to fish. These structures include a large and highly complicated crescent of tissue that sort of curls around the very centre of the brain. It's called the striatum (shown in Figure 1) and it's the main character—the villain—when it comes to addiction. The striatum evolved to select actions that lead to the achievement of goals. In fact, there's not much point to actions that don't lead to goals, so the striatum evolved for the express purpose of connecting actions with goals. The striatum gets a goal on its radar and then sends an action "script" to other parts of the brain (like the motor cortex) to execute, to move the muscles in such a way as to get whatever it is you're after: tapping the keys on your phone, reaching for your wallet, mouthing the words that will get the right response.

But the only way to get an animal to act is to motivate it (unless that action is directly programmed, like the flinching of a worm when you touch it, or the bounce of your leg when your knee gets a tap). Emotion, motivation—these were evolutionary experiments that started around the time mammals first got up off their bellies. If mammals were going to learn from their experiences—unlike their cold-blooded forebears—they would need a more flexible operating system than the *fixed action patterns* responsible for the flinching of the worm and the darting of the frog's tongue. So the part of the brain preserved deep within the cortex—the action brain, the striatum—also became the brain's motivational centre. And it took charge of the one emotion that gets us to pursue our goals, often relentlessly: the emotion of desire.

As the brain's action control headquarters, the striatum has the capacity to produce feelings of attraction and engagement, so crucial for forward-aimed action. But it also gauges the likelihood of meeting that goal, and once that goal is reached, it signals how rewarding (or disappointing) the goal was, compared to expectations. So its job specs include producing the motivation to pursue something, determining how hard you have to try to get that thing, and feeling good (or not so good) about the goal once it is reached.

The striatum learns from experience. It adjusts its wiring according to what felt good in the past and how hard it was to achieve that feeling. *Thus it translates past pleasures into present desires.* When Natalie felt that wash of craving for a shot of heroin, imagining how good it was going to feel, working out what she had to do to get it, her striatum was buzzing. The synaptic networks in her striatum included in the "heroin" pattern (flashing all over her brain) were alive with neuronal transmissions—firing rates well into the red zone, marked "urgent." At the same time, her other goals, like pleasing her boss at the restaurant, getting on better with Grace, and calling her mom, all faded, became insubstantial. The networks that supported those goals dimmed and then turned dark. As with other brain networks, "what fires together wires together," and what doesn't fire together—with the rest of the team—gets left on the sidelines. Those synapses literally lose efficiency and may eventually disappear altogether.

The nucleus accumbens (or just "accumbens" for short) is one of the most ventral (lower) regions of the striatum (as shown in Figure 1). It is tucked deep down in the forebrain, underlying the cortex and limbic system, and it sends its outputs to many other brain parts. In an MRI scan, it shows up as a pair of spots, one in the left brain and one in the right (though it's referred to in the singular: the accumbens). But those spots do a lot of work. The accumbens is a dugout of ancient tissue, and it hums like an engine buried under the moving parts it controls—the other brain regions. The accumbens is the brain part you hear about most when it comes to addiction. It embodies the "hot" end of the striatum, the end with the most connections to other regions that give rise to emotions, expectancies, and actions. To get you to your goal, the accumbens organizes neural activity, not only in the frontal cortex, where meaning is manufactured and expectancies formed, not only in the back of the cortex, where images and memories come alive, but also in the amygdala, which triggers emotional responses, in the brainstem centres that provide the pulp of emotional feeling, and in the premotor and motor cortex, where actions are choreographed and executed. Most of these connections go in both directions. The accumbens gets input *from* these regions. That's how images—needles and powder and the vein in her arm—could influence Natalie's intentions and goals, her scripted pleas to Steve. But it also sends output *to* these regions, including the amygdala, which pulls all sorts of emotions into the mix, and the lower brain parts that could activate sweat glands and wake the butterflies in her stomach. The power of the accumbens to control attention, perception, feeling, and action is immense, and it uses that power to override the ragtag crowd of alternative goals, so that just one goal is left, now glowing with immediacy, pounding with importance. *That's where we're going. That's what we're going to do.*

The accumbens is often considered the brain region where desire is ignited. And most experts agree that intense desire, or *craving*, is the dark horse of addiction. But the accumbens needs fuel to jack up the firing rates of its neurons. And that fuel is dopamine, a neurochemical pumped up from the centre of the brain—the midbrain—located one floor down and a bit further back (also shown in Figure 1). Craving intensifies when the midbrain sends dopamine up to the accumbens. The more dopamine, the more the accumbens is activated, and the more we experience craving. *I just really want it, and I want it right now! I know it's going to feel good, or at least a whole lot better than how I feel without it. And nothing else matters.* Researchers used to talk about this machinery as the "pleasure circuit"—but it turns out that our intense pursuit of goals, like sex, heroin, and chocolate cheesecake, has a lot more to do with desire than with pleasure. Wanting something is not the same as liking something, and most of the accumbens is devoted to *wanting*. Pleasure is a pastry puff, a dessert, a flash in the pan. Desire is what gets us moving, whether that means calling your dealer, driving to the liquor store, or stealing twenty bucks from your aunt's purse. Dopamine—at least where it's absorbed in the striatum—is the fuel of desire, not fun.

A couple of researchers, Kent Berridge and Terry Robinson at the University of Michigan, provided two major upgrades to the neuroscience of addiction. The first was a map of the brain geography of *wanting* versus *liking*. What they found is that most of the striatum is in the business of wanting, and only a small area produces the sensation of liking. It seems that evolution devoted a lot more real estate to desire than to the end state—pleasure or relief—it sometimes achieves. Their second contribution was a formula for the growth of desire for *specific* goals with the rise of addiction, a formula that describes how drugs (and sex, and food, and other attractive things) end up triggering impulsive behaviour. Their work was mostly done with rats and mice, but our brains aren't much different when it comes to the accumbens. The more their rodents were exposed to cues that predicted getting addictive drugs, or even sugar, the more those cues commandeered the accumbens. The cue, the stimulus, whether it was a green light or a horizontal stripe, became increasingly likely to direct the animal's attention and behaviour toward the reward. Berridge and Robinson called this process *incentive sensitization*. And the mechanism responsible was simply the tide of dopamine coursing up from the midbrain to the accumbens. Somewhere between the visual cortex (where the cue was registered as input) and the accumbens (in charge of output), experience altered the wiring, and the cue became the hand that cranked the dopamine pump. In fact, even secondary cues associated with drug-related cues took on that eerie potency. So you could have one cue predicting another cue that predicted getting high, and then the first cue in line—for example, the buzzing of Natalie's cell phone—would start the ball rolling, start the dopamine flowing.

That was Natalie's situation within a month of her unfortunate experiment injecting heroin. She couldn't help the avalanche of excitement, desire, and anxiety triggered by that first hint of drugs on the horizon. This didn't mean she was fated to go and get more heroin. Her actions were not bound and determined by the changes taking place in her brain. But her feelings and thoughts were changing in step. More and more often there came an unrelenting cascade of craving and a narrowing of attention—two streams converging into one—as long as the link between cue and outcome lit up the synaptic networks in her striatum like strings of Christmas bulbs. And, of course, every time those bulbs lit up, it got harder to find the off switch. Every time she introduced a new stimulus to the "heroin" club, like a ringtone for the phone at Steve's parents' house, it got initiated into the fraternity and acquired power over the dopamine pump. Now there were more roads leading to Rome. Now there were more "heroin" synapses linked with each other, so firing rates would climb more quickly, more easily, more predictably, from a greater range of starting points.

That's how Natalie developed a serious habit. It's not that she was *forced* to act on cue—but it became harder and harder to resist the temptation.

~

By the end of May, the fabric of Natalie's and Fred's middle-class life was unravelling. They were simply coasting, not thinking about what they were doing, mesmerized by brain systems far older than the intellectual data banks targeted by their professors. Natalie remembers the slide. She remembers noticing how dour Grace looked when surveying the remains of last night's debauchery— used syringes, overflowing ashtrays—and she'd think, uh-oh, slept late again. And we haven't been cleaning up before going to sleep. Can't even remember going to sleep.

So Natalie wasn't surprised when Grace suggested a change of scene. Since the lease was up anyway. Grace was going to move into a shared house, and Natalie—well, there was no end of options, but for now she'd just stay at Fred's, where she already spent most nights. As soon as they had enough for first and last month's rent, they'd get a bigger place. By the start of fall term for sure. It would all work out. It *was* working out, she told herself.

Not until she received her final grades did she realize how much trouble she was in. She told Fred and Steve this had to stop. She recalls the moment vividly. "I remember looking around my apart- ment, which reeked of stale cigarettes, and being overwhelmed with its total filth; trash everywhere, dirty needles, neglected textbooks. No spoon was left unbent, and no little baggie left unscoured. What made this experience more visceral was my companions . . . [who] were also hopelessly addicted to heroin. It is one thing to be stuck in your head realizing you have a problem, but it is quite another to have two other people validating the truth. We started to talk about it, and then we all cried for what felt like hours. I couldn't wait for it to be over, and the next morning we were all red-eyed, dazed, and sober." Somehow this had to change.

But it didn't change by itself. Circumstances lent a hand.

One evening, a friend came bolting through the open door of their apartment, bellowing that he'd just hit a car. He'd come straight there. He'd obviously been drinking. And he didn't have it in him to face the cops. But he'd been driving Fred's car, and the plates were registered to this address. Sure enough, just a few minutes later, there was an authoritative pounding on the door. Natalie was sitting on the sofa, stuffing pot into a bong (a relatively mild transgression for her), when she looked up to see light sabres piercing the front window. Someone had forgotten to close the curtains.

"We know you have drugs in there," came the steely voice. "Now open up. We have the authority to enter whether you open the door or not."

What they found was laughably meagre: a couple of spoons, burnt on the bottom, classified as paraphernalia, and a single tablet of prescription narcotics. But it was enough for a drug possession charge. So within a week, Natalie found herself staring across a desk at the forehead of an assistant DA, agreeing to comply with a treatment regimen, a plea bargain she was told she could hardly refuse. She pled guilty to the charges and agreed to complete one year of outpatient treatment, after which the conviction would be expunged from her record.

What she hadn't considered was how intolerable the treatment program would become to her. So intolerable that she could not keep herself plugged into its goals. She had endured her withdrawal—not as terrible as she'd anticipated—and stayed abstinent for several months. But she was not a happy camper. She had to attend group therapy sessions three times per week and show up at a local twelve-step meeting three more times per week. It was just too much. All that sitting in a circle and talking about *not* drinking

booze or taking drugs. The other people there belonged to another species. These desiccated drunks constantly sipping the worst
imaginable coffee, desperately trying to hold on to another month
of sobriety. What did she have in common with them? There was
no individual therapy component, no one she felt she could really
talk with. But one afternoon, when the urgings broke through the
surface of her resolve, when she knew she was on her way down,
Natalie pulled the group leader aside after the meeting and almost
pleaded with her: "I really do *want* to want to change," she said. But,
she added in her own mind, I don't want to change into *this*.

So she started using again. She figured out how to fake the regular urine tests by bringing a friend's pee in a bottle in her pocket.
By summer she was shooting up daily. When she was caught the
first time, she got a warning. The second time, she was sent to a
thirty-day residential rehab. They called it a treatment centre, but in
Natalie's view it was nothing more than a patched-up motel run by
a few ex-addicts who figured they knew how to cure people because
they themselves had quit. Her first roommate was a forty-five-year-
old prostitute who would do crack, heroin, whatever came around.
"A nice person," Natalie recalls. "I mean, she definitely had some
kind of personality disorder, but I liked her." Then there was the
guy with the scar. Everyone called him "The Scar." It went from ear
to ear, and people said he'd tried to kill himself with a knife. Natalie
felt sorry for the guy. Long hair, flannel shirt, definitely a loner. This
was his twenty-fifth rehab, he admitted shyly. Natalie thought he
was the most incompetent person she'd ever met. He couldn't even
figure out how to kill himself.

She called Fred a few times a week, but he was in the process of
leaving her. She knew it. When she asked him to bring her something to take the edge off—just one bag!—he hung up on her. He
was moving back to his parents' house, as he'd promised. He was
done. And her future took on a darker hue.

After rehab, she returned to an empty apartment and started working full-time at the restaurant. She continued with the outpatient program, as requested, but also started shooting up regularly again. She'd discovered another way to outfox the urine tests. But the regrowth of this *thing*, this stunted limb, had come about seamlessly, without fanfare, without warning. In her words: "There was no crack to get your fingers under, to peel back that layer of wrongness." The "heroin" circuits fanning out from her striatum to other brain parts were like fibre-optic cables awakened after a brownout. And much of their energy came from anxiety, anxiety about *just being okay.* By now, heroin was both the cause of that anxiety and its only relief. A sense of doom settled over her, a mental and neural pattern that had grown in fits and starts but now oppressed her all day long. She held the hurt and fear inside her while at the restaurant, and then, when she got home, she shot herself up, and those feelings receded like a retreating army, gathering, retrenching, somewhere out of range. For now.

Six months after starting the outpatient program she blew it a third time. She slept late, missed a scheduled group session, and was confronted by her probation officer when she finally pounded up the stairs to the meeting room.

This fiftysomething woman seemed to care at least a little. "I thought you were going to make it," she said. "Come with me."

Natalie followed her into her office.

"You know you're probably going to be kicked out," she added. "Sit here. We're going to search your car. Then we'll figure out what to do with you."

They found a plastic baggie with heroin residue. It was right there in the glove compartment. A clever criminal mind was something Natalie had never achieved. She was finished.

It was an open-and-shut case. She had a public defender, and the best he could do was get her a nine-month sentence. But that wasn't the worst of it. In fact, it was a kind of relief, knowing that it was finally over. The worst part was that her parents would find out. They'd find out *everything*. Her whole family would find out: her aunts and uncles, all those upright Catholics, who never even swore. Who never drank—not even wine at dinner. They'd find out that Natalie was going to jail. And for reasons she never understood, there was only one facility available: a maximum-security prison.

That's where she was taken, directly from court. There was no chance to clean up loose ends. Her dad and sister packed up her apartment several days later. This was it. Two tiers of cells, two beds per cell, and a central observation hub. This was now home.

Almost everyone was there for possession of drugs. Which made it a little less scary. But whether it was because the prison was short-staffed, broke, or just terribly inhumane, the authorities allowed Natalie and the other inmates only two hours a day outside their cells. The rest of the time they had nowhere to go. Time crept along so torturously that Natalie felt she might simply lose it, start screaming, join the crazies. Marking off each square on the calendar was the highlight of her day. Only 122 to go. One down from yesterday. Natalie's first cellmate was actively psychotic, talking to herself, yelling at nobody. But at least she didn't seem dangerous. After she was released or reassigned—Natalie never knew which— cellmate number two showed up: a young, pretty girl who had gotten busted driving through the state with drugs in her car. That was an improvement. "But she was so dumb," Natalie recalls, "she thought Lebanon was some place in America."

To keep herself sane, to begin to imagine a future that she could anticipate without bitter distaste, Natalie started to meditate. She ordered over a dozen books on meditation and mindfulness and read them from cover to cover. They helped. She found that it was possible to just be present in the middle of her chaotic mind, to accept the harsh voices raised in argument there, the agonizing feelings she'd never really let herself feel. She could not deny that she'd been unhappy at the core for many years. And when she opened her eyes and came back to her present circumstances, she asked herself over and over: How did I get this way? What went wrong? Why have I been doing this to myself?

~

She didn't feel she'd been a victim of trauma, a survivor of one of those terrible stories you hear, with disgusting vignettes of physical abuse or sexual abuse, often at the hands of a parent or stepparent. Her upbringing simply hadn't been horrific, hardly even abnormal. Her parents had split up when she was nine, but lots of people came from broken homes without ending up heroin addicts. Her father, with whom she felt most comfortable, wasn't around much after that. Maybe that's because her mom's new husband—it didn't take her long to find one—was such a complete jerk. A grown man who threw tantrums at the slightest provocation. If you didn't fold the laundry quite right, if you were five minutes late for dinner, he'd begin seething with righteous anger. And then, often, he'd explode. But the flare-ups and bullying were never quite bad enough to get her mom to change the situation; she was glad just to have a man around. So Natalie had no choice but to live with her mom and her husband throughout her teenage years. Still, that man—how dare he assume the mantle of stepdad? He was no kind of dad at all.

Natalie had not been happy. In fact, looking back, she realized she'd spent most of her adolescence in a swamp of depression. Her interactions with other kids were always tentative: How did they feel about her? Did they really like her? Yet being at home was another kind of purgatory. So she stayed in her room. It was a habit she'd acquired as an even younger kid, reading books from beginning to end while lying on her bed, not wanting to witness the deterioration of her parents' marriage. And then, as an adolescent, not wanting to be anywhere close to her stepfather. Throughout her teenage years she practiced the art of going vacant, tuning her mind to the static bands between stations, shutting down. She allowed herself to look as miserable as she felt, just in case that would solicit some attention, some help. And she slept a lot. She slept as much as she could.

But she still didn't think her childhood and adolescence had been bad enough to justify her present predicament. Most of the other inmates had awful stories to tell—about all the bad stuff that had happened to them. She couldn't quite believe that emotional suffering is real suffering, that it counts. She'd been trained to see herself as the author of her own misfortunes. Her reveries would conclude with the same refrain: I was never mistreated. It just wasn't that bad.

～

But perhaps it really was. People often take drugs to solve a problem that has no better solution. To alleviate an ache that won't go away, an ache that others may never see, that they themselves can't name. Natalie had found some kind of completion, injecting herself with warmth, with well-being, with a comfort that nobody could deny her. By following a conscientious, almost clinical sequence of steps, she would arrive in a world of contentment that relied on no one—and that no one could take away from her. And it got her out of the house—the house that still existed in her fantasies. It freed her of the nonsensical rules that her stepfather had ingrained in her. Nobody could control Natalie the junkie. Or so she had hoped, so she had imagined. Yet now she found both her body and mind imprisoned, with no chance of escape. At least not until her life was returned to her.

In heroin, Natalie had sought a solution to a problem, intractable since childhood—a problem she was not even aware of. And though that solution never lasted long enough, and it worked less effectively as the months went by, it still soothed her, pleased her, for a little while. But that bit of peace left an enormous wake of longing. And so she came back to claim it again and again. Until it claimed her. Nobody had ever made it make sense, and she hadn't figured it out for herself. There had never been a story Natalie could tell herself that put the pieces together. Until now. Now, on reflection, she could see how they fit.

~

Once she was released from jail Natalie went to live with her mother. Her awful stepfather had finally gotten the boot, and her mom seemed to have a lot more room for Natalie in her life. Cravings for heroin would spike unpredictably for the first few months. But Natalie had taught herself to meditate, and that was sufficient to help her endure them. They would come upon her suddenly and stay for a while, and she would have to fight them or turn away, or else just let them pass—without following them. Then they would dim in intensity and finally leave.

How did she manage that?

We could just call it self-control, bolstered by a nasty set of associations between heroin and all the wrong things that had happened. But it's not that simple. Natalie had to find a self before she could find self-control. She needed the time to reflect, to meditate, to remember and mourn her wounded childhood. All that time in the worst possible environment became the silver lining to a very dark cloud. But the wide avenue of synapses between Natalie's striatum and her prefrontal cortex was a two-way street. And now a lot of the traffic was coming from the other direction. Now her prefrontal cortex (PFC)—the part of the brain that plans, regulates, and controls impulses—began to form new patterns, new habits, based on a more coherent, more conscious sense of who she was. Those patterns became strong enough to survive, to resist disintegration, so they could help stave off the cresting of desire from her striatum. Finally, abstinence itself became a pattern, a habit of its own. Natalie learned that there is a space between craving and doing, and that space could be stretched indefinitely by her own determination, her own commitment to care for herself. Over the next year, the cravings lessened in frequency and finally disappeared.

Natalie has not used heroin since her release from prison. In fact, she's abstained from all opiates, except for one grey afternoon when she downed a few Percocets. The result was not pleasant. "Once the familiar feeling crept over me, I was filled with dread and disgust," she told me, "because it is permanently linked to all the shitty things that happened as a result of my addiction. I cried and cried; it couldn't wear off fast enough." That was it. That was the only reminder she needed.

We could say that Natalie *chose* to stop using drugs, but it's not that simple either. Over and above the conscious, deliberate choice to change her life, the promise offered by heroin had been tainted by too much loneliness, too much remorse, too much anxiety and suffering. So she didn't have to writhe for hours, trying to free herself from the grip of craving. Attraction and repulsion now came in one package. Yet the striatal engines of desire were not extinguished. That would have left a zombie in place of a striving, growing human being. Instead, desire was rerouted. It was now in league with other goals: self-preservation, self-control, a respite from her weariness. As mentioned in Chapter Two, new habits must include remnants of the old—synaptic residue that won't go away. But new habits also transform old habits in unforeseeable ways. For Natalie, heroin no longer meant relief. Now abstinence meant relief.

Had Natalie avoided the chain of events that had landed her in jail, would she have quit anyway? And if so, would it have taken her much longer? Statistics don't throw light on any single case, but most people who become dependent on heroin eventually do quit. Some researchers suggest that, on average, people quit about fifteen years after starting. But a minority of users never quit, and many die, as we know. And even for those who do quit, there's always that disturbing question: how much damage was done along the way?

~

Natalie went back to school soon after her release. Once she'd completed her undergraduate degree, she talked her way into a social work program, partly based on the credentials she'd earned by crawling through the decay of urban America—a place where few college kids ever stray. She got into graduate school, got through the program, and landed a job at one of her placement settings, not only because of where she'd been but because of where she was now. Because she's smart, and personable, and unusually candid about her ghoulish journey. Now Natalie works in an outpatient treatment centre for hard-core drug addicts. And while she tries to help them get on with their lives, she knows, better than most of her colleagues, that there isn't much help she can provide. Until they're ready. Until they've already begun to move on, or at least to imagine moving on. That's when a little nudge can go a long way.

FOUR

The Tunnel of Attention

Brian's Romance with Meth

B rian was jolted from sleep by the sound of a nearby police siren. But it seemed to be winding down as it approached. The
siren wail bottomed out in a low moan. It must be a dream,
he thought. But then a knock at the window shocked him into full
consciousness. He remembers opening his eyes to a concerned face
peering in at him. "Brian, are you all right? Are you drunk?"

"No. Just resting . . . before I get on the road." Peter was a local
cop. Brian had known him for years, not as a friend, exactly, but as
a friendly presence in this suburb of Cape Town, South Africa. He
tried to pitch some energy into his voice, tried to smile.

"Well, you're snoozing behind the wheel. And you're not really
allowed to park here. Just checking."

Brian forced his blood to circulate a little faster. "It's been a long
day, that's all."

The officer disappeared a moment later. But the next thing Brian
noticed was a repeat performance, a déjà vu: the siren winding
down, Peter's face looking through the window, concerned. "Still
sleeping?" he asked, suspicion now in his voice.

"Yeah, just a few more minutes before I get on the road. That's all."

"Brian, you haven't moved from this spot for eight hours. I'm just coming off duty. Look, the sun is rising."

The cop car pulled away, but Brian was left aghast. He remembers feeling completely disoriented. Eight hours sleeping behind the wheel? Had he been so exhausted? Rigid limbs began to explore the possibility of movement. He had to pee. It took several minutes to put it all together. Almost two full weeks of crystal meth, day and night, and he hadn't slept—really slept—the whole time. Maybe a few of those catnaps that came and went like fritzy neon signs. That was all. And now that he'd gotten some of the sleep he so badly needed, what he wanted more than anything was more meth.

~

Brian is one of the sweetest, gentlest people I've had the pleasure to interview while gathering life stories for this book. He's a good-hearted, intelligent man, and he's been a successful entrepreneur since his early twenties (except for one fairly long gap, as you will see). He quit using methamphetamine six years ago. Now his unclouded face shows none of the telltale signs of meth use. His teeth seem to be present and accounted for, approximately the right colour, and his speech relaxed and thoughtful, as though he's got nothing to prove and nothing to hide. Just a few months ago he had his first (luckily mild) heart attack, something he was able to joke about: "I guess my heart had its revenge after all that meth," he told me.

Brian had never taken an intoxicating substance, except alcohol, before age thirty-two. Hard to believe, except that his attention deficit disorder (ADD)—diagnosed in childhood—made it challenging just to think in straight lines. But that was okay. He'd lived inside his brain long enough to know how to focus it, relatively speaking. He could concentrate for hours at a time, as long as the goal remained out in front. In his mid-twenties he'd started a company that sold security systems, and it had flourished in this age of societal upheaval. He was hiring more people, distributing to international clients, putting his company on the map. He was making it happen. But then his life started to become patchy, his goals tattered and confused. His company was no longer the centre of his world, and things with Vera were going downhill fast. When he caught her having sex with another man, he should have tossed the whole thing. They had lost the capacity to talk with each other, and there wasn't much else going on between them. Yet he could not stop wanting her, and wanting her back.

Cocaine helped him to feel more competent, more verbal, more tuned in. It seemed Vera was listening now, and maybe this drug would help him construct new bridges or at least repair old ones. He didn't love the feeling of coke, but he found it useful. It also helped him feel more present, more connected, when talking with his clients. It helped him work well into the night. There was too much work to finish in an eight-hour day, and he couldn't afford to succumb to the floating lassitude and wandering thoughts that came as close as he ever got to depression. But as his tolerance increased, his daily use went over the red line into unaffordable. He had to find another source for the energy he needed to keep ahead of the game.

He started playing around with other stimulants, and landed on a cheap cousin of amphetamine, called "cat" in the local lingo. But the effect was rough and ragged, and it never quite got him past his fatigue. Then he climbed back up the social ladder of the stimulant world and got some methamphetamine—crystal meth— from a colleague: a small plastic baggie of diamond dust, refracting light from the bigger pieces, beautiful just to look at. And a pipe to smoke it in. There were special glass pipes, called "lollies" in South Africa, "which had a status in themselves," he informed me. "Cheap nasty ones or very good fancy ones. . . . Of course I had to have one specially made with a combustion chamber, et cetera." The drug itself was of the highest quality, manufactured by Chinese chemists flown in for just that purpose—or so he was told by his friend and supplier. And indeed there was something exemplary about it, something potent enough to lift him well above the murky cycles of sleeping and waking, fatigue and focus. That was its gift.

At first Brian did not want to admit that he enjoyed the high. He took it because he needed it. There was no other way to stay sharp for fourteen hours a day, and that's just the way it was. He had no serious moral compunctions, but the label *druggie* didn't fit his self-image: a good, solid person, doing his best to get on with life. Still, his relationship with Vera continued to disintegrate. She vacillated between abject guilt and unpredictable rage . . . well, unpredictable at first, until she found a just target: Brian's drug use. Would he have to choose between the two of them? Not likely. He'd already lost Vera. He knew it. "As my relationship was collapsing, my relationship with the substance increased, to fill the gap," he said during one of our interviews. So he bonded with meth. He transferred his sense of attachment from a person to a drug—a drug he could acquire whenever he wanted.

He did like the high. He liked it more each time. And as long as it stayed under control, so what? He would smoke a few hits at about ten in the morning, almost the way other people drank coffee. And then again in the late afternoon. Which created an unforeseen problem: he sometimes couldn't get to sleep. Meth remains active anywhere from eight to eighteen hours, depending on how much you take. So he made sure that his afternoon dose was reasonable. He allotted himself one gram per week, and that kept things under control. He made sure he'd be able to sleep by 2:00 a.m. at the latest. That seemed to work.

Until it didn't. When that surge of clarity, power, and potential rose like a jet in midafternoon, he would not, could not, bail out until he was high in the sky once again. So began the sleep-less nights, strung together for days at a time. Because he could not tolerate coming down, the prerequisite to sleep. Yet fatigue wasn't such a threat when he had the antidote, as much as he needed to beat it back. It was amazing how much you could get done if you went a few days without sleep. As long as you kept the meth level "refreshed"—a term he used often. As long as you kept up what you needed to do, what you needed to be. So he'd smoke again in the evening, and then once again when dawn broke. And really it was one continuous ride, with a few rough spots but no actual breaks. It didn't stop and start. It just went on.

Within two years, Brian was sharing an apartment with a friend whose sole occupation seemed to be selling meth. That was handy, because by now meth had become the centre of his world. From one gram per week he'd graduated to two grams per day. And it soon became clear that the only way he could afford that was to deal to others. So he joined his housemate in this lucrative trade.

But the biggest change in Brian's life was the way its fundamental patterning shifted to some sci-fi scenario—from a rising and falling line that moved through dark and light phases, sleep and wakeful-ness, to some unearthly orbit where day and night were irrelevant, artificial constructions. Meth was like a sun that was always at the centre of his trajectory. It was never further away than the next thought. And living life in orbit changes the way you get from hour to hour and day to day. Wanting it was one thing. Even needing it—yes, he needed it; he'd admit that. But this constant tugging at his thoughts, at the atoms of his attention . . . this went beyond wanting and needing. This was a cognitive mutation. And because he could get it whenever he wanted, the biggest problem was that he usually wanted it right now.

∽

Tomorrow he'd be seeing Megan, and that was the pinnacle of his week. She still had that little-kid cuteness; but now that she was almost ten, her sentences were longer and her thinking more sophisticated. She was growing up, and he didn't want to miss any of it. Despite the separation, he was her one and only dad.

But that was tomorrow. Today was just a regular day. A day that would be spent on the phone, exchanging drugs and money, and in front of the computer, trying to bring his business plans together in a logical sequence. And a meth day. Every day was a meth day. Now that the security business had been sold off, he spent his mental energy working out plans for new, more profitable business dealings: property companies, design studios, recording studios—there was no end of really sharp ideas, though they never seemed to come off. Now he would smoke every few hours, barely coming down from the last hit before it was time for the next one. And of course the sleepless nights made it essential not to wait too long. Because he'd crash, really crash, if he didn't get more meth into his body when the grey tide began to rise. That would be the end of this stretched-out day, this horizonless day.

The night before his visit to Megan, he told himself he had to maintain control. So he *would* sleep tonight, at least for a few hours. He promised himself he would not smoke after 6:00 p.m., which meant he'd be asleep by 2:00, for sure, given his rising fatigue. He had to keep some for tomorrow anyway. Not that he'd smoke while visiting his child—there were some ground rules, after all. But he'd certainly want it when he woke up. And wanting without getting—that was unthinkable, somehow more frightening than anything else. And he might want another hit before the drive—no, not *might*, he *would* want it; who was he kidding?—so that he'd be at his best when he arrived at the house. And after. Especially after seeing Vera and driving away from his kid, the anger and the sadness might start up if he didn't nip it in the bud. So he had to save at least a gram for tomorrow. And now, as night fell and he started to drift, he thought he might as well get the pipe ready for first thing tomorrow. Which meant cleaning it. And then melting a pretty big chunk into the bowl, ready to smoke, so he'd just have to reach for it, get the high going before getting out of bed.

But when it was all ready, grandly ready, sitting there plush and regal, he couldn't resist another hit. A long strong hit. And then he was very high again. Much too high to sleep.

Not that he didn't try. He lay down an hour or two after midnight and sort of pretended. There was no remorse, just resignation, and even an irrepressible sense of triumph he didn't want to confess. He closed his eyes and floated in that lucid state of images and thought particles. He might have slept for a few minutes here or there. He wasn't sure. Then the sun was rising. The clock said six, and the thought of sleep became absurd, almost offensive, despite the wet shrouds of fatigue now dragging him relentlessly downward. Gotta wake up. Seeing my kid in four hours. It's time! And suddenly there was nothing else, no other thought than the imminent kiss of the pipe, just an arm's length away on the bedside table. He had left the lighter in reach too. He'd known this moment would come. Even when he'd lain down to pretend to sleep, he'd known that another hit would be waiting for him in a few hours. And that was all that mattered: that immediate future, calling him, touching him. That was all he'd been able to think about late last night, and now it was here.

Somehow the gram he'd saved for today was half gone. No, it was more than half gone by the time he was ready to leave. He'd just had a nice big hit, true, but how long would it last? So he was on the road and he was thinking: I need to have enough so I can stop obsessing about it every minute. I want my time with Megan to be free and easy. And I really am too tired to go without. Which means circling back to the house to get the rest of the gram. But that wouldn't be enough now. So halfway to Vera's he detoured to the house of the guy he and his friend picked up from—just around the corner, really. He'd be late, but he was often late. In his words: "Everything I was doing had to do with the substance. Even my important relationships, like with my kid . . . I had to score beforehand when I was about to visit my daughter. So a four-hour visit would become a two-hour visit." Megan wouldn't freak out: she was used to it. Then, back in the car, the question was whether to have another hit right now. It was so present, so inviting, in its neat plastic gift wrap. And it would be his last chance before entering the no-fly zone, Vera's territory. But he was already really high, and he didn't want to be too hyper. Being with Megan . . . that should be a mellow time, not laced with breathtaking ideas, talking a mile a minute. And Vera would be watching him until she finally went shopping or whatever it was she did.

Better call her. "Stuck in traffic. Sorry! I'll be there in ten minutes." The silence on the other end told him it didn't wash, but he really couldn't care less. He was flying now. Happy, well stocked, going straight where he wanted to go. To hang out with his kid. Have fun together.

He hesitated before getting out of the car. He wasn't going to take it in with him. Of course not. But he couldn't leave it in full view either. He should stash it in the trunk. But—and this was typical—he was telling himself that he wasn't going to use more until after the visit while, at the same moment, he was busy concealing his topped-up pipe under his cricket jersey, which he heaped cleverly on the front seat, just in case. He could snatch the pipe and smoke it in half a minute. "I'm just going to the car for a sec. Forgot something," he'd say, the words already arranging themselves in preparation for an eventuality he refused to acknowledge consciously. As if his mind were picking up two distinct channels simultaneously.

And a third channel, more like a narrator's voice, intoned that it didn't matter, because he was the boss and he could handle it. The contradictions meant nothing. It was always a high-wire act. The contradictions were just tricks that made everything more exciting.

$\sim$

The striatum, especially the accumbens, doesn't act on its own. No brain structure functions autonomously. Rather, the accumbens connects with more than a dozen other major brain structures, and two of its most reliable allies are the amygdala and the orbitofrontal cortex (OFC), both of which are close neighbours (and both shown in Figure 1). These three lords of the emotional realm work as a team, but it's the accumbens that ends up taking the ball and scoring. The accumbens is selected for action, for turning motivation into movement. Its two henchmen set up the play.

The amygdala is a small structure embedded beneath the cortical layers on each side of the head. The amygdala is *sub*cortical: in other words, it's a more primitive system (around the same vintage as the striatum, hundreds of millions of years old), and it's responsible for the emotional spray paint that stains every important experience, almost instantly, with emotional tone and colour: the fear you suddenly feel when you hear footsteps behind you on a dark street, or the jolt of shame you experience when you spill coffee all over your pants, or the sudden excitement and pleasure that Brian felt as soon as "crystal" came to mind. Whether it came from his visual system, because he'd suddenly discerned the pipe on the bedside table, or from his memory system, where thousands of meth associations swam about like fish in an aquarium, the image of "crystal" was immediately imbued with emotional meaning by his amygdala, whose synapses had been sculpted over two years of continuous use. Feeling and focus go together for the amygdala, especially when it comes to directing your gaze to the source of the good, bad, or ugly. Amygdalar emotion corrals many brain systems with a simple but fundamental command: *Pay attention!*

But the OFC is far more sophisticated when it comes to directing the play—organizing diverse cognitive agents into a mental phalanx, a mind with a mission. The OFC joins the amygdala within fractions of a second, picking up and amplifying its emotional beam. The OFC is a patch of tissue on the bottom surface of the prefrontal cortex; its job is to connect emotion to expectancies and fashion a rough action plan—both (relatively primitive) functions of the prefrontal cortex. The OFC holds information about how nice and gratifying some things are, especially now, given present circumstances, and how miserable and repugnant other things are. And it can combine these contradictory judgements, as it did for Brian that day, based on the complexities of the situation at hand. Brian's OFC reacted to the high-voltage message from his amygdala, breaking it down into a more nuanced interpretation. If his OFC could talk, it might say, *This is very good, despite the chance of getting caught. Especially when I have to see Vera and feel on top of things. Especially when I haven't slept for days. It's coming my way, and that's what I want.*

The OFC is the amygdala's more discerning cousin, a bridge between the limbic system and the prefrontal cortex, providing more shadings, more detail, and clear implications for action. Nerve bundles from Brian's OFC to the rest of his prefrontal cortex amplified every facet of attention, rumination, memory, and anticipation—the *where, when,* and *how much* needed for the next high, based on the last high, how much was left, how hard it would be to get more, and how high he wanted to get. That's how the images of meth that infested Brian's life gave rise to unique emotional arrays, potent yet particular, while simultaneously capturing his thoughts. That's how, over time, through the cultivation of diverse fields of synapses, he wound up being unable to stop obsessing about methamphetamine.

The accumbens gets its information both from the direct beam of the amygdala and from the nuanced expectations of the OFC. That's when the accumbens goes into its predatory stance, its back arched and its claws extended, with only the goal in mind, only desire in its circuits. But at the same time the accumbens encourages its two henchmen to share its feast of dopamine: the dopamine fountain activated by the drug images themselves. Streams of dopamine now flow outward from the midbrain to fuel all three systems, augmenting their firing rates, confirming their alliance, and marshalling the thoughts that rise to consciousness. Planning, sneaking, subterfuge, self-deception, scanning for opportunities, averting risks—these are more cognitive than emotional, and they're based on activation of parts of the prefrontal cortex that resonate to signals from the OFC. All this while quivering in anticipation of another hit. Should he keep driving to Vera's? Should he circle back home for the rest of his gram and be an extra twenty minutes late? Or should he stop by his dealer's house and buy another? Each of these cognitive gambits sprang from his OFC's massive investment, right now, at this moment, in the value, meaning, and availability of the next hit of crystal meth.

But are these processes abnormal? Does all this bouncing around of neural activation suggest the gaunt profile of a diseased brain? Actually, no. In fact, what we see is a brain that is in the throes of strong emotion, while making its way in a social world characterized by conflict and contradiction. (A fairly typical world, especially for addicts.) We are designed to connect the many components of cognition with feeling itself. The triumvirate of the amygdala, OFC, and accumbens—the motivational core of the brain—evolved precisely for the purpose of linking cognition with emotion, thought with feeling, and then putting the best available plan into action.

Compare that with the brain of the field mouse, constructed to beam visual attention in synch with its drives: fear, hunger, and sex. Mice need to be highly vigilant when larger animals, like cats, are suddenly spotted or smelled, which is why their beady little eyes remain transfixed on whatever it was that just moved in the grass. For that, an amygdala and a primitive accumbens are all you need. But for us great apes, brains require a good deal more. They need to fashion strategies, conscious or not, to deal with the emotional demands and opportunities of the moment, based on long-standing memories and the nuances of the present situation, nuances that continually seed and prune the OFC's connections with its neighbours. That's how the human brain is *supposed* to work.

The OFC is the bottom floor of the most sophisticated structure in the primate brain, the prefrontal cortex (PFC). In that role it deserves considerable credit for translating the raw glare of emotion into thoughts, expectancies, and a readiness to compare potential outcomes, advocating some and rejecting others. This requires the OFC to recruit its more advanced cortical neighbours, those dedicated more to reflection than passion. That's its job! And if that reflection turns into rumination, we can't suddenly look for some foreign entity, point our finger, and shout, "Disease!" Rather, we need to realize that the emotional stakes are extremely high, and competing goals have been relegated to the sidelines. Rumination is the result of a normal brain, doing what it's designed to do, when the brain's owner has entered a cycle of seeking and finding the same thing over and over again. Whether that something is chess, tennis, or methamphetamine, it becomes more mesmerizing, its alternatives less interesting, with each cycle. That's motivated repetition. It's exactly what goes on in the mind and the brain when we are in love, or at least infatuated, and we cannot think about anything or anyone other than the person we desire.

But there's a catch that ensnares addicts just a little more than everyone else. Just enough to break the tie in the tug-of-war between attraction and willpower. It's called *delay discounting*: the tendency for humans, other mammals, and even birds to value immediate rewards over long-term benefits. Delayed rewards are discounted. Their value is reduced. Delayed negative consequences are also discounted—in other words, delayed punishments seem less severe than immediate ones. Typically, delay discounting is studied by conditioning lab animals or humans to expect two or more rewards, each at a different delay interval. The one expected sooner is designed to be objectively less valuable (less food or sucrose for rats, less money for humans) than the one expected after a delay. Yet we earthly creatures dive for the short-term gain, despite the overall reduction in good things acquired. I'll refer to delay discounting as "now appeal" for the rest of this book.

For addicts, long-term benefits are self-evident. They include happier, healthier relationships, physical health, money in the bank, self-respect, and the likelihood of staying out of jail. They are moulded into the slogans chanted at twelve-step meetings: *Keep coming back, it works if you work it. This too shall pass.* And they include the avoidance of long-term suffering, the inevitable descent into misery forewarned by messages from the War on Drugs, or into cancer and death, as threatened by dire words and images on cigarette boxes all over the world. Yet all these future rewards (and probable disasters) are minimized, dissipated, stripped of value, by the intrusive, glittering promise of the immediate goal. Which is why Brian veered off the road leading to his daughter, leading to the rest of his life, and found himself on his dealer's doorstep instead.

Not surprisingly, the accumbens and its neighbours are to blame for now appeal—an unfortunate bias in the firmware we've been evolving for aeons. Like the slouching that results in backaches—a painful spin-off of the achievement of upright locomotion—now appeal is an evolutionary side effect. Yet how could it be otherwise? The accumbens evolved to get the animal to go for low-hanging fruit, available sexual partners, whatever is most accessible—a habit it has kept to this day. Dopamine rises with anticipation, rushing in to rev up the accumbens, when rewards are just around the corner. So dopamine (in the striatum) is, once again, the villain here. Its hypnotic attraction to immediate goodies distorts the perspective we could have (otherwise) achieved using our more advanced cognitive abilities. We are so familiar with this built-in bias that we'd be surprised and disappointed if Hollywood lovers didn't rush into each other's arms with stunning velocity. We lunge for the immediate. Which makes life tricky for lovers and addicts both.

With now appeal, the pattern of activated synapses contracts to a narrow beam, focused on the immediate. And a lot of that narrowing is anchored in the network of roads between the accumbens and the OFC. So it's no surprise that research subjects lying in an fMRI scanner show increased activation in both the accumbens and the OFC when they choose an immediate payoff over a later reward. In real life, these two partners in crime focus your attention on the attractive traveller smiling at you from the next table, not the comforting hug of your spouse waiting at home. And Brian's ruminations were repeatedly pulled by the shimmering proximity of the next rush of meth, not the warm visit he would later have with his daughter. Because the next hit of meth seemed *so* valuable. It's not that his brain was working improperly; it's just that he'd arranged his life around a single goal, an inanimate lover, and his brain did what brains do—it revised itself accordingly. It composed a hierarchy of goals in which Vera, even Megan, were no longer on top.

Addicts are excessively now-oriented, more prone to delay discounting than the population average.[1] But nobody knows quite why. Perhaps it's a personality characteristic they've shown since childhood, putting them at greater risk of addiction to begin with. Now appeal is highly correlated with an impulsive personality style, hardly a disease in itself, but a well-known forerunner of addiction problems. Unluckily for Brian, ADD diagnoses overlap with both impulsivity and now appeal, so the cards were already stacked against him. It's also likely that now appeal gets more severe when life is lived on the edge and all possible rewards and punishments are partitioned into two categories—score or go without, now or never. But neuroscience offers us another important clue: now appeal becomes augmented—that is, immediate rewards become even more attractive—when the most advanced (dorsal) regions of the prefrontal cortex are disturbed by a magnetic field produced in the lab. I'll describe this research in a later chapter. But for now, the important point is that researchers have observed a disconnect between the dorsal PFC and the striatum (or related regions) with addiction as well as now appeal. In both cases, the diminished capacity for perspective taking and self-control has a concrete neural parallel.

∼

Brian was finally where he belonged that morning: sitting on the sofa next to Megan, who still cuddled into him like a small child as they watched an old movie on the TV. And while the continuous roar of the meth drowned out emotions such as guilt and remorse, he couldn't help thinking about what had just happened. He hadn't been able to wait. That was it. He could have looked forward to this happy oasis all morning long and spared himself the crazy gyrations: the detour to his dealer's house and the frantic hit he'd taken as soon as he returned to his car. And it was so unnecessary. He was so high at this point he could not really focus on the movie, on Megan, or anything else. His thoughts were running too fast. He hadn't needed another hit that soon. And yet he hadn't been able to wait. He'd felt that the extra half hour to get to Vera's, and the twenty minutes he'd need to extricate himself before she left them alone, would be too much to endure. The meth was just five minutes away, said his own voice. A left turn, another mile down the road. And there it would be.

As so often happened, immediacy closed the deal. That's what wrenched the steering wheel to the left, seemingly independent of his own rational reticence. And now, just a couple of hours later, sitting here with his kid, picturing him, Megan, and Vera at the upcoming dinner in the next room, the broken family temporarily held together by wishful thinking, it called to him again. It was less than a minute away, on the front seat of the car, where he'd left it.

~

The day finally came when Brian's housemate and fellow dealer, Gordon, got into trouble with the local gangs in Cape Town. Gordon was Brian's friend but also the head of their dealing business, the guy who kept his head clear enough to think things through—until he got into an argument with the leader of a highly ambitious gang who demanded monthly payments in return for staying away. One night the gang leader pulled a knife on Gordon. Gordon then pulled a gun out and shot him. Brian hadn't known that Gordon even possessed a gun. From that night on, Gordon was out of the picture, hiding somewhere, and Brian ended up alone in the apartment, trying to keep his head down and use his natural charisma to keep other gang members from retaliating. After years of doing whatever he could for Gordon, and now immensely worried about him, he pretended to distance himself from his former ally.

But it didn't last. It couldn't last. Four armed men burst through the door a few weeks later. It was time for Brian to return the money he supposedly owed. They started yelling at him and pushing him around the room. Then one of them tried to hit him with a nearby broomstick. Brian blocked it. When the man came at him a second time, he blocked the blow again and the broomstick broke. Another guy picked up an electric guitar, Brian's most prized possession, and swung it at him. This time the attempted block broke his arm. He was in big trouble, but still thoughts of meth invaded. Should he pretend to go along with them? If he agreed to their demands, he'd get another hit of meth right now. They'd sit around and enjoy it together. Otherwise, they would clean out everything he had in the house. Was he really thinking about getting high at a time like this? He continued to amaze himself.

What he ended up doing was equally foolish, but probably the only chance he had to save himself: he jumped out of the window. It was a second-storey window, and he landed hard on a garage roof. Then, fuelled by adrenalin, fear, and indignation, he ran from roof to roof until he spotted a car right beneath him. He jumped onto it, denting the hood profoundly, then found that he was surrounded by shoppers and salespeople, some sympathetic, some terrified. The car owner was approaching. He could see her reaching into her purse, surely to call the police. The gang members would be there soon. There was nothing to do but run.

He convinced a doctor friend to load him up with morphine and construct a splint. He was permitted to stay the night. Then, in the following few days, the pandemonium of his life began to settle into sluggish uniformity, a time to do nothing at all.

Within days he made his home in the garage of another friend. It was a strange arrangement: the friend, someone Brian had been kind to in the past, shooed him out every morning and locked him in every night. That way his wife would never find this drug-addicted derelict residing on their property. But it worked. He spent his days wandering the streets after selling his car for next to nothing. He attached himself to a familiar crowd of druggies, and most of his time was spent on the lookout for a hit of meth he couldn't afford, "tongue hanging out," as he recalls. He sometimes got rewarded. He had helped a lot of these people when he was on top, bailing them out of jail, lending them money for a lawyer or for drugs. He had earned a measure of respect in this ramshackle community, but that halo wore thin as the months went by. His fall from grace was a long, slow mudslide. He got his drugs with money from odd jobs, or he went without: wishing, hoping, begging for another act of kindness.

The desperation for drugs crackled through every waking hour. He hated going without, but he hated the relentless gravitational pull on his mood and his meaning so much more. Eventually his friend Joseph, who'd given up meth a couple of years before, began working on him, trying to convince him to come with him to a meeting of Narcotics Anonymous (NA).

"Look, things will get better," Brian insisted. "All I have to do is find a job."

"You're not thinking straight," Joseph replied. "You just want money for more drugs. You never really stop, do you? You need a rest."

Intermittently over the following months, Joseph would ask, "You still using?"

"Nah, just having a hard time for a few days." In Brian's mind, lying was no more immoral than the truth. And strangely, bizarrely, he did not consider his main problem to be addiction, as seemed obvious to everyone else. He still longed for Vera; he still saw meth as a substitute for Vera. That was the fracture that could not be fixed.

Finally he went with Joseph to an NA meeting. He sat down with this group of bedraggled strangers. At first he could hardly focus on their stories, their strange slogans and metaphors, their efforts to gain some perspective on their struggles and their needs.

And then he began to notice that their stories sounded something like his. He saw parallels. He began to realize that the word "addict" just might apply to him after all. He went to more meetings. And that was when his internal struggle burst wide open. Meth was all he had to look forward to. It was all he had left. But he had to leave it behind. He had to leave behind this thing that was so much a part of him. All addicts feel that jolt of loss when they finally, reluctantly, turn their back on this friend, this lover, this part of themselves.

It's agonizing.

Brian never particularly liked the rituals and the dogma of NA, but he saw that it was helping him. Here in this gathering he didn't have to pretend to be something he wasn't. He was accepted as the tortured person he had truly become. That was soothing to him. It relieved his loneliness. He also started to accept that Vera was long gone. And because his need of her and his desire for meth were so intertwined, it seemed they could be laid in the same grave, put to rest at the same time.

～

For the next four and a half years, Brian lived in a shack on the property of someone else who wanted to repay a past favour. At the beginning of that period he would smoke meth occasionally, but it seemed to have lost its sparkle. It was part of another life now—not his life. It felt like a hindrance rather than a boon. He continued to attend the occasional meeting, but most of the work he needed to do he did on his own—or with a little help. For six months during that period he made weekly sojourns to a psychotherapist, whose job it was to gather and hold the unrecognized shards of his life— particularly his early life—until he could put them together and make sense of his style of interacting with the world. Early in the therapy he saw that the way he approached every situation was an expression of habit, not conscious reflection. Then, over the following months, he began to discover the nature and origin of the habits that defined him. His mother had been a "screamer," as he put it, a highly volatile woman who would flip between angry criticism and exaggerated concern for his well-being. "I was never good enough," he remembers. "Neither was my brother. But I was not even allowed to swim until I was thirteen years old." He simply expected other people to act the same way. He wanted to be free of their expectations; he was convinced that a stormy rebuff lay waiting around the next bend. He wanted to be his own man, but he also needed reassurance, almost unremittingly—or at least more than Vera could provide.

Like Natalie, Brian had to fashion a story, a narrative, to explain himself to himself. And a crucial chapter in that story was the link between his parent's failures and his own. The narrative he unearthed was painful but necessary—a foundation for the growth now under way, the transformation he willed himself to continue, day by day by day.

As for drugs, there were rare occasions when he still smoked meth, not very big hits, but still . . . To his surprise, these lapses no longer led to an urge for repetition, nor an orgy of obsession and rumination. In fact, they led to almost nothing. Rather than overvalue the meth—or devalue his own future—he would muse: That wasn't the brightest idea. Now I won't be able to sleep tonight. So what had become of now appeal, the lure of the immediate? Perhaps Brian could now imagine a future self valuable enough to pursue—a future self that was a continuation of the thoughtful, insightful self he was now becoming. A self that formed an unbroken path from his childhood to the rest of his life. And because he'd stopped using regularly, the disconnect between the dorsal regions of his prefrontal cortex and his striatum could reverse, reknit, regrow. Perspective could regain its foothold and get stronger with time.

Meth lost its relevance for Brian. And because his own habits no longer fertilized the synaptic pathways he'd laid down all these years, those pathways gradually faded away. The seductive value of meth was expunged. The growth of new foliage began to obscure the trail he'd been following for so long, until that trail was hardly noticeable. Now the forest spread in every direction, revealing pathways he'd never seen before. Exuberant growth, novel possibilities. There was so much to think about. And to do.

Brian repaid the owner of the property by taking care of his animals for several months. He became a shepherd by day, and he spent the rest of his time reading everything he could get his hands on. About relationships, about how they worked, when they worked, how they failed. And about addiction. Instead of seeing himself as a classic victim of failed relationships, he began to imagine that he was the kind of person who could fix them, for others as well as himself. Perhaps caring for animals, shepherding sheep, and reading about rudderless lives converged into a sense of what therapy might offer. Pieces of a puzzle seemed to come together with insight, luck, and time. From the straying cats of his own thoughts to the straying unfortunates of his community, Brian emerged as someone who could bring order to chaos, gather the lost ones and help them find their way home. This image of himself has grown from a daydream to a burning commitment over the years since then.

For a year and a half, Brian ran a therapeutic community for addicts and alcoholics, right at the farm where he'd cared for the owner's sheep. Now, back in Cape Town, he runs a funded outpatient facility that provides free services for dispossessed and addicted wanderers—a club to which he recently enough belonged. Over the past three years he's completed a postgraduate diploma program in addiction care. And recently, right around the time of our first interview, Brian was accepted into a master's program in addiction and mental health. That master's degree will help him continue his work; it will add nicely to the cluster of achievements posted on the wall of his office and the internal bulletin board of his renewed self-image. A new relationship with a smart and beautiful woman, also in the community mental health field, rounds out this chapter of Brian's life. He says this one is different. He's in love but he's not desperate. He finds that he can be passionate without being crazy, and that life without drugs is anything but boring.

~

The brain evolved to pursue goals by focusing attention and motivation on likely sources of pleasure or relief, especially those right in front of our noses. Delay discounting—now appeal—is simply a label for the shortsighted nature of this highly focused state. Yet its outcome, the shrinking of perspective and funnelling of desire, is an evolutionary adaptation that has allowed us to survive and thrive, maximizing opportunities for enhancing our lives and those of our offspring. This feature of our neural nature is not only normal. It's essential—for a species that needs to think fast and act fast, in order to survive and prosper in a world of multiple options.

By pursuing the same goal repeatedly, as we often do when we find something or someone we constantly desire, we strengthen synaptic networks in the parts of the brain that underpin motivation and focus, parts that fashion long-standing representations of what is valuable and important. This is learning, of course. But the repetitive quest for the same goal, always shining more brightly than other, more distant goals, accelerates the learning process. The more attention narrows, the more specific—and limited—the learning becomes. Which only narrows attention more the next time around. The feedback cycle becomes a shrinking spiral, a cone contracting to a point.

The brain is a habit-forming machine. In addiction, as in love, our habits can grow with remarkable speed. Brian's attraction to meth began to accelerate exponentially when his relationship with Vera disintegrated. In just a few months it morphed from a lifestyle habit to a full-blown addiction. But the plasticity lost by the brain as habits cohere and crystallize is never completely lost. Except in the case of extreme organic damage, or when environments have become unnaturally restricting, the brain always retains enough plasticity for new pathways to grow. They may not grow quickly while addicts are first trying to quit, because a tightly focused beam of desire for a well-defined goal has not yet converged. Instead they are driven by more abstract desires: for freedom, novelty, and lasting contentment. When those goals finally do converge and become concrete, as they eventually did for Brian, new synaptic pathways sprout vigorously again—beyond the well-worn routes that once determined life's narrow boundaries.

FIVE

Donna's Secret Identity

Donna felt herself receding, a little more each month, from the people in her life. Even her moments of connection, with her mother, her brother, and most of all Michael, began to feel contrived, a game whose rules she was starting to forget. She told me, during one of our first interviews, that the distance suited her, because nobody could have understood how important the pills had become—and that wasn't going to change just now. She had accepted that, seeing it as fair and logical in its own way. But it cleaved the world into two domains, the public and the private—the *very* private.

These last three years had been so stressful, with Michael out of commission half the time, lying on the sofa, nursing his spine. With her work more demanding than ever, talking to the families of kids with killer diseases, surfing that kind of despair and staying on her feet. The nurses considered her a saint. And she *was* a saint. Except for that one little weakness, the one she saved up for the end of her shift. And that ability to save it—that proved she was still in control, didn't it?

This is how she remembers the week it all fell apart:

She'd left work early, this particular Wednesday, and was now on her way to a family party, a house full of relatives she had no wish to see. Howard, Michael's first cousin, would be arriving right around now. She'd planned everything methodically, as usual. But her anxiety increased steadily. Her mother-in-law, Gert, from whom she'd become galactically distant, was hosting the event. She couldn't shake the feeling that Gert knew a lot more than she let on. Sure, she kept her smile on high, and Donna reciprocated with her own smile, equally fake, making amends. For what? She wasn't certain. But there were too many signs that all was not well: family members acting strangely toward her—overly accommodating. And of course she had been acting strangely—how could they not notice?—leaving the room and coming back ten minutes later. Then leaving again. Going to the bathroom . . . again, Donna? Entering the house by the back door and then announcing, after who knew how long: Hi! I'm here!

How could they not know?

Yet no one had accused her, no doors had slammed in her face; Michael was still . . . the same old Michael. Bottom line, no one had stopped her. And that was the strangest thing of all, because it made it okay. It let the dream continue. The little girl went unpunished. She must not be as naughty, as *bad,* as it sometimes seemed.

Donna has an unusually clear recollection of her internal dialogue at the time. She can tick off the details of her inner life with precision. What she describes is a familiar debate that went on in disconnected snatches. That afternoon she drove along the 405, which runs the length of L.A. Cars lined up before her, exchanging lanes in zigzag patterns. She tried to control her anxiety by focusing on the immediate, balancing her will between the gas and the brakes. But when she got there, would she go for it? Would she make a beeline for Howard's toilet kit, where the odds of finding some pharmaceutical goodies were actually pretty high? He seemed like a guy with a few health problems: pudgy, middle-aged, well medicated. She felt certain she would find something, if she could get in and out without being noticed. The wrongness of it didn't matter anymore. Right? Wrong? The question was increasingly abstract.

By now she'd accepted what she'd become. There wasn't a nice word for it—she was a thief. She only stole drugs. But still, she *stole* them. And she lied about it afterward. So she'd become a liar too. A year before, she'd attended the funeral of a friend's mother. Once at the reception, she hadn't hesitated to ransack the medicine chest of the deceased woman. She won't need *these* anymore, Donna told herself, to siphon off the shame. But then her friend appeared, just as she shut the bathroom door behind her. "The one downstairs was busy," Donna remembers saying. And her friend seemed not to notice the bulges in her purse and the anxiety in her eyes. Donna felt the familiar flood of relief: all's well that ends well.

Then came her string of thefts from Marsha, who lived in the condo below theirs. Marsha had shared some of her painkillers with Donna and sold her the whole supply a few times. But Donna had got in the habit of stealing most of the leftovers—Marsha must have known. Yet nothing had happened, and an eerie sense of invulnerability arose from these close calls. Hide and seek: home free! Waltzing back to her front yard, adrenalin pumping.

She was gradually creating a new self-image. In her words, "I became another person: I became a liar." Donna the liar, Donna the thief, Donna the druggie, Donna who takes what she needs. Ever since childhood she'd tried to be a good girl. Her work at the hospital was based on helping others. But maybe she really was evil. Maybe a well-deserved tragedy waited just around the corner. Or maybe she was free to wander for some time yet, through this netherworld of moral relativity. She shrugged mentally whenever the debate surfaced enough to annoy her. It was not a problem she was presently able to solve.

∾

There were a dozen people in the house when she arrived, and more on the way. Most sat around the living room, soft drinks in hand, chattering about this and that. They were all relatives and family friends, getting reacquainted after months or years apart. Howard, the newcomer, was onstage for now. She said hello. A few heads turned, smiled, nodded. But only her sister-in-law, Cathy, shouted a greeting in return. The others: she could feel their eyes on the back of her head as she hung up her jacket.

She walked up the hall, into the kitchen, prepared to help Gert. But the older woman told her to just go and relax. Everything would be ready soon.

Donna can't recall any reason for her heightened anxiety that day. It certainly wasn't the first time she'd plotted a pill heist. She remembers feeling excluded, and she remembers thinking of it as a mixed blessing. People's eyes seemed to slide past her, as though she occupied a blind spot. She imagined a card taped to her forehead: *Drug Addict*. Everyone else could read it. And then look away. Yet the trappings of warmth were comforting: the babble of inanities that rose and fell with its own rhythm. It was just a family party. Another family party. And there were more coming up, more guests arriving each day, anticipating the grand finale: Michael's fiftieth-birthday party, early next week.

She decided it was time. She told herself that the warmth in this room wasn't for her. She'd find her own warmth.

∾

She'd been finding her own warmth for most of her life. Donna had grown up in a family that barely functioned. Her father had been sexually abused by his mother, so he claimed, and it's not the sort of thing people make up. Then he'd had his first adult sexual experience at the age of thirty-five, with Donna's mother. It was never clear to Donna what kept them together after that. Dad had been an alcoholic for as long as she could remember. And suicidal. As a young girl, she was compelled by the need to take care of him. And so was the rest of the family. He lived in paralyzed immobility, and somehow that was everybody's problem.

Donna remembers her mother's constant admonishments: Don't be dramatic, Donna. Let's just keep quiet, Donna. No need to make a fuss. In fact, very little emotion of any kind was permitted. The important thing was to keep Dad alive, to keep the ship from sinking. There was little warmth. Little laughter. And certainly no anger. Any expression that could be interpreted as resistant or defiant sent silent alarms screaming. Would Dad feel thwarted? Would he feel attacked? Or trapped? If so, all hell would break loose. Spilled milk at the local restaurant was apocalyptic. It was her brother's doing, but nobody was allowed to go out to eat for two years after that event. Even when she was a baby, Donna's crying was muted. Her mother would turn off the baby monitor if she cried too long—something she admitted to her daughter years later. Dad must not be disturbed.

Instead of being cared for as a child, Donna became a caretaker. "I just figured out this was my way of being in the family: being good. I got a lot of credit for that," she told me. "I was really self-contained, reading books early on, not playing with other kids." In fact, she didn't feel comfortable with other kids: "I definitely had the feeling that I didn't belong . . . that there was something wrong with me." But within the family she was precious. Her father confided in her by the time she was eight. He told her way too much about his personal demons. And she made his salvation her personal mission. Dad was into cars, so Donna got into cars. Dad did not touch.

So she tried to get close to him through conversation, talking about cars or sports or the stock market, topics borrowed off the shelf in the living room. But worst of all, Dad often seemed on the verge of suicide. So Donna got out of her bed and went to check that he was still breathing, several times a night, surreptitiously tiptoeing from room to room, just as she would silently creep through her in-laws' house twenty years later.

Donna's need to hide her feelings took on new forms in adolescence: new shoots fed by her maturing vitality, whole new branches reaching upward. At age twelve she began starving herself. She told everyone she'd become a vegetarian. But her purpose was to sharpen the blade of self-control until she could turn it outward or inward at will. She didn't take it to the extreme of anorexia. Rather, she was admired for being thin and healthy. That's how good she was. Vigilant and determined, the guardian of her own emotions. By sixteen, she was competent, pretty, liked by everyone, but suppressing anger as a matter of course. Once she received her college degree, she got a competitive job in a hospital, working with the families of very sick children. There was no end of praise for her strength and capacity to nurture, and no recognition of the hidden resentments that might easily reverse the generosity pump.

She discovered drugs in her mid-twenties: Vicodin, prescribed for a neck injury she'd got from a bicycle tour. At first it was just a painkiller. It surprises her, looking back on that period of her life, that opiates held no particular appeal for about a year. Then she tried doubling the dose, taking four at a time, and that's when they began to reveal their more intimate gifts. They allowed her to hide, remain invulnerable, and feel loved, from the inside, all in a single package. The baby monitor could be turned off for good. She no longer needed rescue.

But the reservoir of warmth began to run dry. It had to be refilled. She learned to forge her own prescriptions—daring but efficient. A year later, after a couple of close calls, she got scared and stopped. She found out what withdrawal symptoms felt like: the dripping eyes and nose, the sheen of overstimulation, the cramps. She stayed off the pills for a year and a half, and during that time she met Michael. They began to live together; they got engaged, then married. But Michael came with a dowry of opiates. His body was in far worse shape than hers had ever been, with several disks in his lower spine slowly disintegrating. After the wedding, he retreated to the sofa, having hurt himself moving furniture around their new apartment. He got Vicodin by the truckload. So Donna started to help herself.

Her life continued to drift in a direction she did not choose, like the movement of tectonic plates, invisible until the crisis. She loved Michael; at least she thought she did. But Michael's spine kept worsening. Now liquid morphine and OxyContin were lying around the apartment. She took what she liked. She developed a special fondness for OxyContin, but she would take any opiate she could get her hands on—Vicodin, Percocet, Dilaudid—wherever she could find it. Even friends' medicine chests and unwatched purses. But Michael's supply was her insurance policy. And against all odds, he seemed not to notice.

Except once when he caught her with her hand in the cookie jar. He seemed annoyed rather than angry. So she admitted to a fondness for opiates and promised to see a therapist, to work things out. She didn't keep that promise, but she became more careful, and that was the end of the matter.

Her hunger for drugs continued to mushroom, and by necessity she became a schemer, building on lessons learned as a child. Self-satisfaction was now inextricably linked with defiance. She savoured each triumph, almost trilling with glee when she pulled off something particularly daring. She found additional sources: a new doctor, friends who were willing to share or sell their drugs or look the other way when she stole them. And relatives—an especially rich vein. Her in-laws had pill bottles stuffed in half a dozen drawers: kitchen drawers, bedside drawers, and of course the medicine chest in the bathroom. She knew she'd become addicted, more psychologically than physically. Yet she felt she could handle it. She was still in control. She didn't take drugs at work. She saved them up until after her shift, by which time she deserved to get high. She was still the responsible health-care worker, taking care of others' needs. But she was determined to take care of her own now as well.

~

Donna described the next scene in agonizing detail. It seems to be etched in her mind, like the events that precede any major trauma, physical or emotional. She had checked all the bathroom drawers, the medicine chest, many times, over the previous weeks. She had checked the closets, the drawers of the bedside tables. She knew there were no more pill bottles worth her efforts anywhere in this house. Unless Cousin Howard had brought his own.

His suitcase would be in the guest room. A shaft of determination filled the hollow space in her stomach, just as it always did before she made her move. She exchanged a few pleasantries with those around her, then excused herself: "I have to make a phone call. I'll be right back."

She walked past the kitchen, exuding nonchalance. But it seemed to take forever to get to the room at the end of the hall. She stepped inside and closed the door.

Howard's suitcase lay on the floor beside the bed. She sat on the bed, leaned over, unzipped it, and picked up the phone in one sequence. She pulled out items of clothing and piled them on the floor, until she saw the vinyl sheen of a toilet kit. She pulled it out, slowly, carefully, and placed it on the floor, next to the pairs of socks that tumbled out with it. Damn! She'd have to put everything back exactly as she'd found it. But first things first. Her right hand pulled the zipper open and began to dig, while her left held the phone that would be her excuse, if anyone—

And then the door opened. Just then. Howard stood at the threshold for a frozen moment, looking directly at her right hand.

"What are you doing?" he said. "What the fuck are you doing?" His voice rose with every word. "Are you crazy?" He lurched into the room.

"I—"

"Why are you going through my suitcase?" His voice settled into an accusation, righteous but still baffled.

She couldn't speak.

"Is it money? Is that what you're looking for? Here, have some money!" He was shouting now. The door had closed, mercifully, but here was this man yelling at her.

"Here! How much do you need?" He dug bills out of his wallet and threw them at her.

"I—" She felt paralyzed.

"You what?!" Droplets of spit landed on her face, and then she looked down at the mess she'd made. The zipper of the toilet kit was open. For a moment, a helpless longing rose in her throat. If only I'd had two more minutes. And then the world came off its hinges: she suddenly saw that things would never be the same again. Never again, from this moment on.

Familiar excuses rose like bubbles, but they burst before they got to the surface. There was no way out. Nothing she could say would work this time.

"I have a drug problem," she said softly. There, it was out. Her hands began to shake, adrenalin pumping. But this didn't feel quite like fear. It felt like shock, slapping her; and shame, gathering now in little pools.

"I have a drug problem," she repeated, tasting it. "I was looking for drugs."

Howard just stared at her, uncomprehending.

"My husband knows," she added, and hearing her own voice pleading for leniency, she began to see herself as others would see her. Because Michael *didn't* know. Not really. Not the extent of it. Not the desperate need, the thefts, the lies. But now he *would* know. And so would everyone else.

∼

In previous chapters I described the neurocircuitry of desire, from its origins in the midbrain and accumbens to the OFC—the orbitofrontal cortex—where value and expectancy are forged from raw emotion. Donna's OFC was a veteran where drugs were concerned. Its responses were practiced, even stereotyped. For several years now, its synapses had generated a glow of anticipation whenever drugs were nearby. But the signal didn't stop there. When drugs were on the horizon, OFC activation continued flowing upward into the higher reaches of her prefrontal cortex (toward the top of the head—dorsally). Each time she thought about, dreamed about, or obtained pharmaceutical opiates, each time she sat back in that familiar cloud of comfort, the stream of activation travelled to these higher prefrontal regions, modifying connections with every wave.

Over multiple occasions, the anticipated relief afforded by drugs (or other addictive goals) continues to etch patterns in the OFC. The OFC is a glutton for emotional learning, so it is subject to synaptic changes from infancy onward. But the changes now taking place in Donna's brain continued northward, along a grand boulevard of nerve fibres. One of the regions along this path lies on either side of the midline or fissure that divides the two cortical hemispheres, like strips of farmland along a river valley. This region is called the medial prefrontal cortex (medial PFC), and, because of its location, between the swampy terrain of the OFC, at the bottom of the PFC, and the drier headlands of reason and judgement near the top, this region becomes dedicated to a task both cognitive and emotional: understanding oneself and other people. In this way it resembles the learning that occurs, not in infancy and early childhood, but in middle childhood, when the brain encodes new constellations of social knowledge.

The medial PFC is the core of the social brain, where interpersonal reality is sorted into two fundamental entities: self and other. The medial PFC is where we make sense of others' actions, translate them and interpret them as to intentions and goals, and evaluate our own actions and goals in response. Here we formulate our perspectives, our prejudices, attempting to connect with those we admire or love and distance ourselves from those we distrust. Psychologists believe that we come to understand the inner lives of other people by playing out their intentions and motivations on the instrument of the self, imagining what it would be like if this were *me*. At the same time, as developing children, we formulate our sense of self by borrowing, combining, and revising the characteristics we perceive or imagine in others. That's how we design our identities, by the age of eight or nine, and how we redesign them as we continue to develop in adolescence and adulthood.

Along with a region at the rear corner of the temporal lobe (the temporal-parietal junction), the medial PFC becomes activated when people think about others' characteristics and intentions *and* about their own characteristics and intentions. In fact, the medial PFC is part of a much larger web that comes alive when we imagine, daydream, and rehearse possible interactions with others. But the medial PFC is particularly important for connecting our self-image with our emotional goals. We define ourselves in synch with those goals, and that self-definition gets cemented by the strengthening of synaptic connections between the medial PFC and other regions.

This network gets reconfigured as teenagers solve their challenging social problems, tweaking their identities while shopping
for shoes. And for Donna it became reconfigured once again with
drugs, revising her sense of self, her values, and her view of other
people. Drugs were allies that provided the warmth her parents had
withheld. They were big news in Donna's emotional world. So the
pathways created by her attraction to drugs modified her identity:
she began to see herself as a taker, not just a giver, a controller, not
just a victim—at least when drugs were around. These aspects of
her identity had already begun to emerge in adolescence, fertilized
by her teenage experiments at controlling herself and others, but
they had remained mostly hidden behind her "good girl" persona.
Now they bloomed more fully in the half-light of her addictive
thinking—a branch growing from the trunk of her personality at an
odd angle, deformed but resilient.

The medial PFC is activated when we judge ourselves, adjust
ourselves, become ourselves. So it's not surprising that it gets reconfigured by the repetition of an experience with immense social and
emotional meaning. That experience might be falling in love, either
with a lover or a child, or breaking the law, or joining a religious
sect. Or it might be drug use, since drugs directly stimulate core
bodily sensations and usually perturb interpersonal relationships.
Experiences of this sort are the agents of developmental change in
identity, morality, even personality. But such experiences need not
be outlandish to influence development, and they have nothing to
do with disease.

Waves of desire shaped Donna's medial PFC into a matrix supporting two self-portraits: Donna the benefactor, who took care of
others, fulfilled their needs, and repressed her own, and Donna the
defiant, who took care of herself because no one else would. "It became a double life," she told me. "I was still a good person, still taking care of people, still successful. . . . People would be shocked to

find out about this other me." She slipped back and forth between these two personae, each barely acknowledging the existence of the other. Not quite a split personality. Rather, a kind of chameleon road trip. She still wanted to be accepted and loved, but it was no longer the only game in town. She continued to practice her warm smiles and generous offerings. Yet when drugs were available, when that singular stream of activation geysered up her prefrontal midline, she could put all that aside. And instead find satisfaction, even triumph, in deceiving those around her, and getting away with it— a new solution to a very old problem.

~

Now, under Cousin Howard's accusatory glare, the two configurations of Donna's identity collided and burst. Donna the druggie had never been hauled out of the closet, onto the stage of familial scrutiny. Her druggie self could not be seen, could never be reconciled with the conventional morals of her in-laws, her siblings and cousins, her parents, and her husband—especially her husband. So now she would suffer. The gush of shame and fear came precisely from the tear in the wall that had kept those worlds apart. If they found out she was a drug addict there would be no forgiveness. She had taken them all for a ride.

For two days Donna went through the motions of her life. Nobody had reacted when she came back into the living room. They must have heard the yelling, and yet nothing changed. Nothing changed when she got home. Michael was his calm, solid self. Except that he seemed preoccupied. He wasn't quite present. But that was a relief, wasn't it? He didn't turn on her, disgorging contempt. Nor was *she* present in the usual way. She wasn't *with* him. But she watched him, very carefully, waiting, like a dog waits with shoulders hunched for a sign of anger on the face of its owner. Would they tell him? Had they told him? Did they even know?

For the first time in months, perhaps years, she needed desperately to hold on to the people in her life. All other goals—including the drugs themselves—quickly faded to grey. The meaning she got from opiates began to lose its distinctive quality. While waiting for the verdict, she began to forget what it was the drugs had meant to her and why they had meant so much. She imagined herself wandering, a bent-over cripple, a pariah, through a world empty of human warmth. She dreamt that her key no longer unlocked the front door. And when she woke up, she told herself this was real. She might lose it all. She stood at the edge of a chasm, looking out at a future world in which her family scorned her, where she had no job, no friends, and no Michael. She pictured all this, many times over, while waiting for the drama to play out.

~

There was another party on Friday evening. Howard was there, but nothing came up. He was cordial with her. And she found herself hoping that maybe, just maybe, this thing would blow over, and she would return to the strange underground life she'd been living. Few words passed between her and Michael on the drive home, until he said with uncanny calmness that Howard had invited him to brunch tomorrow. She didn't breathe. She tried to read the silence.

Late next morning, Donna was at Cathy's house, co-hosting a baby shower. Nothing could be more mundane, more normal, and this calmed her. But Michael and Howard were together, at brunch, and whatever transpired there would dictate the course of her life. She checked her phone every ten minutes. No calls. She wished that God or someone would intervene this one time and let it be okay. Let Howard keep his mouth shut. Let Michael not find out.

When Michael finally did call, she listened intently for clues. He asked when she'd be home. His voice sounded neutral and her hope brightened, a sputtering flame.

Until the moment she walked through the door and the flame died. Michael was sitting in an easy chair in the living room, curtains drawn, the room semi-dark, with darker shadows at his feet.

"Come in and sit down," he said without expression.

She sat.

"I know you've been using," he continued. "I know what happened with Cousin Howard."

Her body froze. She continued to stare at him.

"My parents have known for weeks. They caught you on video—"

"What?"

"They set up a video camera in the corner of the bedroom. You wouldn't have seen it. But it saw you . . . going in and out of the bathroom . . . so many times," he said, his voice finally rising from its monotone. "Can you tell me what you were doing in my parents' bedroom and bathroom?" He stared at her without hostility, but with a mixture of awe and contempt that was far more unsettling. As if he didn't know her well and didn't want to know her at all.

So she told him. In a quiet, deliberate voice, recounting the incremental steps she took when she was hunting for drugs, not finding them, feeling the disappointment, the resentment, and then craving them all the more. She told him exactly what it was like when she imagined she might have missed something. An unnoticed drawer, the cupboard under the sink. And he listened carefully, nodding his head from time to time, because her record was accurate. It matched what his parents said they'd seen on the video, detail by ugly detail.

From this correspondence between her words and his nods, she knew that the video must truly exist. But that was unthinkable. She imagined what the camera must have seen. Donna raking through drawer after drawer, again and again, like a crazy person. The shame doubled her over: a kick perfectly aimed at her stomach.

"They're going to show it to me," Michael said.

"No, Michael! Please don't watch it! Don't ever watch it!" she howled. Because she was sure he would stop loving her if he were ever to witness her desperate hunt, like that of a starving animal.

"It doesn't matter," he said, his interrogation nearly finished. "You're going to call your parents and you're going to go into treatment. Right now. If you don't, I'm going to leave you."

And there it was, finally. Her worst fear, the shock of ice-cold water.

She crumpled to the floor, the last of her defences, the last of her strength, completely sapped. Terror washed over her in waves, each one followed by a sucking backwash of shame. Her body jerked with the thrust of these emotions. Her knees folded into her chest. Her eyes closed tightly. She could not bear to see him looking at her. Not now, not ever.

Yet beneath the storm was a small room where another voice spoke. The game is over, it said. You can let go now. You don't have to struggle anymore.

"I'll do whatever you want," she said out loud, eyes still closed.

For the first time since childhood, she was completely helpless. She *would* do whatever she was told. There remained no other motive, no other plan. And uncannily, whenever the pounding shame let up for a moment, she felt what she least expected: relief. Huge relief. The surrender of a self held together too long, with too much effort—trying to be the person everyone thought she was, yet determined to get what she deserved—now giving way to exhaustion.

∾

There was a lot of crying that night. A lot of phone calls, gut-wrenching apologies, messages of forgiveness and hope. Calls to Michael's parents, then to her own parents. Everyone crying and loving her, even forgiving her. That was the most surprising thing,

Donna recalls: the compassion and care that came from her mother, and her brother, when what she expected was rejection and scorn.

That night they left a message on the answering service of an addiction specialist. She called back with an appointment for Monday at 9:00 a.m. That left two days of limbo. Donna hardly moved. She lay on the floor for much of the time. She would not permit herself to sleep in a bed. It was one way to punish herself. She wanted to die, she remembers. She felt she did not deserve to be alive. But she had neither the energy nor the determination to kill herself. She hardly spoke with Michael. She later discovered that she'd been placed on suicide watch, with Michael checking on her every hour or two. Yet despite the gusts of anguish, there was peace here. She was dazed, numbed, often feeling very little, or feeling the relief of no longer having to try to be anyone at all. She and Michael maintained this holding pattern, waiting for Monday.

When Monday morning finally came, Michael and Donna's brother half carried her to the car. She had never been in treatment of any kind. She then met the woman she would continue to talk with for several years. Someone who had also been an addict once upon a time. The therapist wanted her in a residential setting as soon as possible. Donna fought it. She didn't want to go. But she thought of Michael, waiting in the waiting room. She finally agreed, giving up the last shreds of what seemed like independence.

Two days later she was admitted to a thirty-day program that offered group and individual therapy. Patients were taken to twelve-step meetings several times a week. Donna's first few outings were to an AA group, but she asked to be shifted to NA (Narcotics Anonymous), and that suited her better.

The thirty days passed. Then Donna moved in with her parents. That was a period of mixed emotions, but there seemed far more warmth in that house than she remembered. After two and a half months, she went back home to Michael. During this whole period she continued to meet with the therapist she had first seen the Monday following her "breakdown"—what else could she call it? She explored the strange new world of total honesty. She liked this person, and she experienced another first: the sense of being understood.

She began to discover the motivations that had brought her back to the pills time after time: the safe feeling bestowed by the drug and the thrill of defiance, even vengeance, that went along with it. Both sentiments were gratifying, but in very different ways. She discovered that, for as long as she could remember, she'd believed that anyone who knew the real her wouldn't want anything to do with her. And she began to feel the anger she had gathered in response.

~

Her therapist wanted to hear the whole story.

The urge to delete her inner world had emerged long before her drug-taking days. The equation that guided her life was that she must hide her feelings of need and vulnerability, her disappointment, her anger, in order to be accepted, tolerated, loved. First as a child, in a family configured around her father's fragility. Then, years later, as a young woman, enmeshed in relationships with abusive men, one after another. She remembers one particularly horrific relationship in which she allowed herself to be tied up with a gun to her head, in order to satisfy the man she was with. Sacrifice was the key to acceptance. As she put it, there was "the moment where my *self* just left."

Now she sees the similarity between abusive relationships and drug addiction, at least as it played out in her own life. In both cases, she had to give up a lot if she wanted to get something. She says she "would walk out of relationships physically broken, just like people who walk into rehab." But the impulse to repeat these disastrous encounters, with men or with drugs, was exactly the same. In between relationships, she would starve herself, not eat for long periods. The important thing was to deny her actual needs, to suffer that deprivation, until she could no longer stand the emptiness.

After her neck injury had healed almost completely, Donna discovered that Vicodin offered something besides pain relief: a sense of being cocooned in warmth—without having to cut off her feelings, without having to hide from herself. But of course she could not share this discovery. Hiding her new obsession from others wasn't easy, but it was not nearly as hurtful as hiding her needs from herself. In fact, the hiding itself became satisfying; it released flaring sunspots of defiance, moments of triumph, in a life otherwise ruled by compromise. It worked, as long as she got the intimacy she needed from her drugs. For Donna, as for Natalie, a certain fusion took place—not the result of the drug itself, but the result of a collage of historical and present wounds, and the fit between what was missing and what was gained.

With her therapist, Donna replayed the messages she'd recorded as a child. "So much was done with silence in my family. Which is what I repeated with other people. I was very withholding with Michael, then defiant through the secrecy." Her therapist held up the mirror, and Donna realized she'd been biting back depression for nearly her whole life. This shocked her. She'd never seen herself as depressed, but now it was obvious. She had denied so much of her pain and chosen self-control and self-abuse instead, with her parents, with men, and with drugs. The important thing had always been to stay hidden. That was the one rule she'd never broken. Until now, sitting in this room with someone she hardly knew.

~

Donna remained off drugs for fourteen months. Then came a period of taking tramadol, a mild opiate, sporadically for four to five

months. I'm not sure how she obtained it, but her use was motivated by her emotional state, not by medical necessity. Like Brian,
she did not quit overnight. Yet Donna's "relapse" may have nurtured
a larger cycle of growth. She and Michael separated as soon as he
found out about it. He cleared all his belongings out of their condo
within two weeks. But this time she did not fall apart. She didn't
even come close. This time there was only relief. Michael's reserved
manner and emotional distance were not what she wanted in a life
partner, but leaving him had never crossed her mind. Wasn't she
the needy one? Didn't he deserve her undying loyalty? Now she
wonders out loud whether her return to drugs was an unconscious
means of breaking up a marriage that wasn't working. There's nothing tidy about becoming addicted, and nothing tidy about quitting.

Yet quitting was relatively easy the second time around. She realized that the NA meetings she'd been attending felt stifling and
moralistic, at least to her, and there were other ways to move on
with her life. She made conscious choices to rekindle relationships
with her friends, and she found most of them to be tremendously
supportive of the process she was going through. They recognized
that she had a choice to make, but they didn't draw a line in the
sand. They didn't threaten to cut her out of their lives if she strayed
from her own resolve. She quit her job at the hospital, though she
continues to see patients privately. The hospital job was another obligation she'd felt compelled to maintain, until she realized it just
wasn't for her.

At the time of our last contact, Donna had been abstinent for
two years. She sounded confident and enthusiastic about her new
life. Not only was she drug-free; she was free of a job and a marriage
that had cast her as a selfless provider. That was a role she'd played
since childhood, a role she'd rebelled against by stealing tokens of
warmth from those around her. She now sees that her needs, her
moods, her selfish moments, and her sometimes volatile nature

aren't toxic to others. Nobody is dropping dead around her. In fact, her interpersonal world is more spontaneous, more loving, than it's ever been. And no doubt her medial prefrontal cortex is changing yet again. Synaptic bundles long kept apart are finally knitting together to create a single cord, flexible and strong. Donna continues to see her therapist for the occasional tune-up. She's not completely out of the woods. But it seems she's in pretty good shape.

I asked her if she felt she'd grown as an individual during her addiction and recovery. She laughed and told me that she's never felt so strong, so happy. Donna made it obvious that not only is addiction a developmental journey, but it's a journey that continues through the period of recovery. In fact, by the time I'd finished my interviews with Donna, the term "recovery" no longer made sense to me. "Recovery" implies going backward, becoming normal again. And it's a reasonable term if you consider addiction a disease. But many of the addicts I've spoken with—including Donna—see themselves as having moved forward, not backward, once they quit, or even while they were quitting. They often find they've become far more aware and self-directed than the person they were before their addiction. There's no easy way to explain this direction of change with the medical terminology of disease and recovery. Instead of recovering, it seems that addicts keep growing, as does anyone who overcomes their difficulties through deliberation and insight.

SIX

Johnny Needs a Drink

One day, near the end, Johnny woke up on the tile floor of the kitchen. The imprint of his face would remain there for weeks, he told me. "It was like the Shroud of Turin." At first he couldn't tell if it was day or night: the light was grey and everything was vague and floating. Then he really came to. He sat and thought for a while. Finally, he hoisted himself up, went to get his sleeping pills, dumped them on one side of the table, put the rum on the other side, and said out loud: "Suit yourself, take one or the other. If you choose the Bacardi, keep it under control. Otherwise, take enough pills to kill yourself. Get it over with." He sat there for several hours, waiting for the sun to come up. Then he sat there for another long while, the alcohol gradually wearing off, wondering what time it was. He finally got it into his head to turn on the TV. It was 11:00 a.m.

The problem was that Johnny couldn't keep it under control. For nearly six months, every day had followed exactly the same pattern. He did not leave his apartment, except for occasional shopping runs. He no longer went in to work. His job was now drinking. He had to drink as soon as he woke up. He might wake up at 3:00

p.m. or at 3:00 a.m., but whenever consciousness asserted itself, his very first thought, his very first requirement, was alcohol. He would wake up, go straight to the fridge, pour himself a large glass of Bacardi, add ice, then some Coke. But the ratio of Coke to rum shrank with each passing week. He could gulp a whole tumbler at once. At least the first one. The next would go down a little more slowly. The call of the alcohol was so strong that he would go to the fridge immediately on waking even if he had to pee. He couldn't wait. He would pour his drink, take a slug, and then carry it with him to the bathroom. The glass had to be in his hand from the first minute of consciousness to the last. But that was a span of only four hours.

~

Johnny is now in his late sixties. I plied him with questions about this period of his life. I've heard many stories from people whose addictions reached the red zone, but I was astonished by the extremes he described. He had lost all connection with time and place. He had constructed his own personal hell, and that's where he lived. I wanted to understand what it was like, in as much detail as possible. He obliged me. He didn't mince words. He spoke slowly, thoughtfully, deliberately, without prompts. It felt to me like he was still surprised, or at least bemused, by the fact that he was alive at all. He didn't wait for my questions, although he answered them when they came up. Otherwise, his story came out in one continuous stream.

~

When the rum hit his stomach, its first effect was to begin to settle the nausea. At least that's how he explained it to me initially. But in a later interview, Johnny recalled that he didn't always wake up with nausea. What he awoke with each time was the anticipation of nausea—joined by the anticipation of overwhelming anxiety. Before he had a chance to check either prediction, the rum was already on its way down his throat. The first half hour was a period of desperation giving way to relief. The next hour was pleasant, enjoyable. He still liked the feeling. He felt relaxed, carefree. This was the pot of gold. Then, for the next two and a half hours, he became increasingly sedated. It was no longer pleasurable. Nor did he expect it to be.

So why continue drinking?

Johnny described this last phase of his four-hour day as "abusive"—which clearly meant self-abusive. In his words: "I would tell myself, you know you're going to end up in the same terrible place tomorrow. . . . So I would drink at that point because I didn't want to face myself, didn't want to be rational." Johnny had terrible thoughts during those final two or three hours, drowning in sedation. He knew he was digging a hole that he would not be able to climb out of, that would lead to his own death. He pictured his business falling apart around him. He'd been trying to take care of things by phone, without ever going into the office. But it wasn't working. And his employees would lose everything too. That wasn't right. In fact, it was unforgivable. He'd say to himself out loud: "It won't be long before you can't stay awake any longer, before you can't remain on this stool." Then he would pour himself another drink. It seemed the only thing he could do. Sometimes he lay down in a fetal position. He would often have to crawl on his hands and knees to bed.

It was just him and booze. That was all there was. He careened between astonishment and disgust at the idea that he still couldn't—or didn't—stop. "I used to pray for relief," he told me. "That I would choke on my own vomit. I didn't want to be here anymore." So why didn't he commit suicide? Because he didn't possess that kind of determination and courage, or so he believed. Further grounds for disgust. Rather, he would pray: "Oh please, God, don't let me wake up tomorrow."

~

Johnny spent much of his childhood praying, though he did not completely trust the god he was praying to, and certainly didn't trust the priests who guided his prayers. His father ran a hardware store in a town in Ireland, where he and his five siblings were raised in relative comfort. The business continued to thrive, and it was a matter of pride and honour that his brothers and he were given the opportunity to go to a private Catholic school. Johnny's turn came at the age of eleven. It was a boarding school, run by priests, but Johnny wasn't eager to be cut off from his home and his friends. Yet to pass up the opportunity was unthinkable. It would undermine every principle his parents lived by. He wished to please them, not hurt them. So he went.

His bed-wetting was still a problem at home, but it got worse as soon as he moved into the dormitory. Of course the presence of other boys was a source of tremendous anxiety, and that fuelled the bed-wetting. Yet each boy was given a cubicle off a central corridor, providing a little bit of privacy. That was a blessing. Despite some logistical problems, Johnny was able to deal with his own wet sheets without being noticed. But it was also a curse. The separation of the cubicles made it easy for the priests to visit certain boys privately, without being observed—and to abuse them sexually. This was common knowledge among the boys, yet nobody talked about it. The boys would not discuss it with each other, and their parents simply wouldn't believe it. Or, if they did believe it, as Johnny suspected his parents might, it would destroy their faith in the Church, the mainstay of their spiritual lives. Johnny seems to have learned about the abuse from his older brother, who'd been at the school several years when Johnny arrived. His brother had been a victim, and this information—however it was acquired—tilted the world off its axis for years to come.

Johnny rocked himself to sleep at night, as he had throughout his childhood. It was the only way to soften the brittle surface of his fear. Maybe that's why they left him alone. The rocking—and the bed-wetting too. He didn't seem quite normal. But he heard whispers and other sounds as he rocked. The gaunt shadows of the priests creeping about. He tried to pretend it wasn't what it was. He rocked and rocked.

Johnny describes his father as his best friend in childhood. When he died, Johnny was twenty-one, and there was little left to keep him at home. He had become a rebellious young man, partly in response to an older brother who now ruled their home like a private fiefdom. His mother receded, his other siblings each found a way to resist the new monarch, and six months later Johnny left for England.

Many Irish teens made a pact with God not to drink alcohol. The emblem of the agreement was a Heart of Jesus, which was engraved on Johnny's tie clip at boarding school. He didn't mind sticking with this commitment. Alcohol meant very little to him, and he wanted to be on good terms with God. But after a while, living the life of a young man in England, he felt awkward sipping his lemonade when the others were drinking beer. So he graduated to shandies—beer mixed with lemonade. It was no big deal for him. Not a formidable line to cross. He didn't like the flavour much, but he found that he liked the effect.

Johnny hadn't yet recognized the anxiety and depression he'd lived with since childhood. It wasn't until much later that he reflected on the causes of his attraction to alcohol. For years he'd been committed to achieve, to be successful—it didn't matter at what. He set high standards, as his father had. But he almost never felt he had attained them. Then, in his mid-twenties, he discovered a way to free himself from anxiety. After three or four drinks, the anxiety just vanished, not to return until the following morning. Here was the feedback loop that would bring him to his knees years later: anxiety, relief, then loss and longing, round and round, digging ruts in the fertile soil of his striatum, laying down pathways to his midbrain and back. Along those pathways, springs of dopamine pooled—released in a flash flood when it was time to drink again.

It took Johnny about four years to shift from normal drinking to serious drinking, as he calls it. During this period, the cues that pointed to the pub, after work or after a game, activated fields of synapses—grown heartier with each repetition—that stood for something different. Alcohol became a symbol, the core of a network that included a promise of peace, cessation of stress, relaxation. Neurally and psychologically, it invaded and overtook the companionship network, like crabgrass colonizing a lawn. For example, around the age of thirty, Johnny would drink to calm his nerves so that he could be relaxed and sociable with the clients of his firm. He'd make sure to order a bottle of wine if the client and his wife were to join him for dinner, and he wouldn't talk much until after he'd had a glass. But within a few years he'd order a second bottle halfway through dinner, knowing that he would be drinking most of it. Or he'd wait until he'd called them a taxi and said goodnight, and then order the second bottle for himself. Now the social interaction was a conduit to drinking, more than the other way around.

Johnny got married and divorced twice before his final showdown with alcohol. He raised children, moved from an employee to a manager, made it in the retail world, and established security for his children. He did all he could to be a responsible provider. But relationships didn't come easily. He didn't take criticism well, and he felt resentful and helpless when his second wife seemed to want to control him. He recalls that her declarations of love were closely followed by a string of commands: what he should wear, where he should sit, when he could go out and when he must return. He admits that he has a particular distaste for "being told," maybe a lineage of reactions that began with the priests, strengthened by the autocratic rule of his brother. But the death of this marriage hit him pretty hard. This was the marriage that should have lasted, and Johnny gave up something irreplaceable when it ended. He lost close contact with two teenage daughters, whom he loved very much. He might not have known quite know how to show it, except by writing cheques. But still . . . He became a man living alone.

His new apartment gave him what he wanted: freedom to do exactly as he pleased. Which included more drinking. He was sixty-one years old. About time he was his own boss. But the novelty soon wore off, and Johnny fell into the habit of going to a local pub to socialize. He liked the guys who hung out there, he often bought rounds, and he gradually became one of them. That saved him from having to cook his own meals. It also saved him from a chafing boredom, a sense of irrelevance, ghosts of his failure to be the kind of husband and father he'd idealized. The pub became a beacon. He would tell himself that he wasn't planning to go on a particular night and change his mind five minutes later. Or he might sit at home in front of the TV for an hour and suddenly tell himself, this is boring. Then off to the pub he went.

Of course, the pub was most attractive, not because of the company or the food alone, but because of the particular way these human needs interlaced with his need for alcohol. The sense of a home away from home and the liquid that muted his restlessness converged at the pub. And that was an enticing mix. At this stage, Johnny's desire for drink was still what you'd call impulsive. It would pop up suddenly, in connection with a particular time or place. There was energy to it. It was still frivolous in a way. There was some joy in it. He thought of himself as a social drinker. And maybe he was stretching the definition. But it was nothing like the next phase of his drinking, the compulsive phase, which waited around the corner.

For the next two and a half years, Johnny was a regular at the pub. But the amount he drank increased steadily, until it became the occasion for shame. By the final few months, he was unable to get home by himself. One of the boys would help him find his way and keep him from stumbling off the sidewalk. He got to the point of having no memory of the night before. And the others would tease him. "Hey, Johnny, you weren't your usual gentlemanly self last night. Groping Madge like that . . . " It wasn't true. He was sure he'd done no such thing. And then he'd verify that he was just being teased. But he couldn't help discern an element of reluctance, even aversion, in the men he drank with. Yes, they cared about him. They were worried about him. But really, getting that pissed every single night?

They would try to help: "Give it a rest, Johnny. Time to go now." But more and more often he'd fall asleep right there at the table. And he'd leave the pub only when someone shook him awake: "Come on, let's go home." And then there were the neighbours to worry about. Now and then he'd get pulled out of the bushes by someone else coming up the walk. And that was disgraceful. It bothered him a great deal.

The shame of being an out-and-out drunk drove him to a new routine. He felt there'd be less shame involved, maybe none at all, if he did his drinking at home. So he'd stop at a supermarket on his way home from work and get a few bottles. And some supplies to go with them. He didn't need much. Rum and Coke had become his drink of choice. Fast, efficient, went down easily. Once he got in the door he had to have one right away. And then another. And then another.

Within a month or two he stopped going into work altogether. He tried to take care of his business by phone. His journeys out of the apartment became increasingly rare. There were five stores where he could buy rum, and he'd pick up some food at the same time. He didn't like going to the same store repeatedly. He had his dignity to protect. Then he pretty much stopped going out at all.

He was drunk all the time, and he couldn't trust himself outside the walls of his home. He got his employees to bring him what he needed. "Are you going to the shop? Would you mind picking up six bottles of wine and a couple of litre bottles of Bacardi?" They felt obliged. They felt they couldn't say no, because he was paying their wages. He didn't want to think about it, but the shame seeped in.

Drinking had finally become compulsive. He did not seem to be able to resist it. And when Johnny turned that corner, night and day, morning and evening finally lost all meaning. The beam that drew him to his rum snapped on like a perversely powerful electromagnet as soon as he realized he was awake. And it stayed on throughout his waking hours, until the power source, his own consciousness, finally dimmed and went out.

≈

I've characterized the striatum as the source of desire, of craving. Remember, the striatum is a spiral of tissue near the centre of the brain that first identifies an action to be pursued, a goal to be attained, and then generates the motivation to go after it. But the striatum is complicated. It's comprised of several components, which evolved to take charge of different kinds of action and the different kinds of motivation needed to achieve them. Quick, sure actions, uncertain actions, automatic reactions, and actions that are still under construction. Most of the striatum absorbs dopamine as its fuel, but dopamine engenders different neural responses, depending on where in the striatum it's sent and what kind of receptors consume it.

We can divide the striatum into a bottom half or southern hemisphere—the ventral striatum, which includes the famous accumbens—and a northern hemisphere, the dorsal striatum, which works quite differently. (The brain sketch in Figure 1 gives a rough impression of the positioning of these structures.) As described earlier, the accumbens is the source of impulsive action, the feeling of suddenly wanting to do something without much thought about consequences. The accumbens is the front-runner of addiction because it is highly attuned to the perceived value of the goal. It is oriented to rewards; and drugs, sex, booze, and gambling are all about rewards. The accumbens is the centre of the whirlpool that nearly drowned Natalie and Brian. And it rapidly forms connections with its close neighbours: the amygdala, which triggers emotional responses, and the OFC, where the value of goals is registered. I've called this network the "motivational core" of the brain. But the northern hemisphere of the striatum, the dorsal striatum, has a different mode of operation. It doesn't care about the value of rewards, it isn't attuned to likely pleasures, and it's not about to send its owner on an impulsive hunt for goodies. Rather, the dorsal striatum records connections between stimuli and responses, so that well-learned actions will be linked with particular stimuli (i.e., cues) and get triggered by those stimuli whenever they are perceived.

Most of us learned about stimulus-response (S-R) learning in Psych 101. Pavlov's dog got famous for salivating at the sound of a bell, a response that became automatic when the bell rang before supper. No forethought was required. But salivation may not be the best example of this kind of learning. In real life, cue-triggered responses are often behaviours rather than glandular responses. For dogs, such behaviours include running to the door when you put on your coat or jiggle your keys. For humans, they cover a broad range of rapid-fire behaviours, like stepping on the brakes when you see taillights up ahead, and locking the car when you get home.

Especially when those behaviours are annoying or difficult to erase, they are considered compulsive, not impulsive. Because you feel you just need to do the thing that's linked with the stimulus, without any expectation of the outcome. If there's any expectation of how good it's going to feel, it's secondary. That feeling is not what drives the behaviour. Rather, there's an urge toward an action, a very particular action, and it's hard to turn off that urge until the action is completed. Obsessive-compulsive disorder (OCD) is the hallmark of unbridled compulsion, and a lot of the brain changes in addiction parallel the neural signature of OCD: both include activation of the dorsal striatum when learned cues are perceived. What's important to realize is that many goal-directed actions start off impulsive, but when they're repeated for weeks, months, or years, they end up as compulsions.

I've highlighted impulsivity in the biographies so far, but Brian's use of methamphetamine had strong compulsive features, as did Donna's pursuit of pills, which showed the desperation that can mark compulsive acts. And Natalie began to use heroin compulsively, without forethought, a month or two after she started shooting up. In general, synaptic changes in the striatum—propelled by repeated experiences—go north over time, from ventral to dorsal territory, and that's pretty much where they end up.

A closer look at the brain shows a remarkable pattern of neuroplasticity, one that seems both beautiful and dangerous. When the accumbens draws dopamine up from the midbrain to fuel its goal-seeking impulses, it also sends messages back to the midbrain, causing changes in the pathways that carry the dopamine. As time goes on, the midbrain sends dopamine to more northern portions of the ventral striatum, which signal the midbrain to send dopamine even further north, and eventually as far north as the dorsal summit. This process works something like an upside-down maypole. A spiral of dopamine pathways is woven along the curving spine of the striatum, from south to north, each pathway overlapping and interleaving the previous pathways, which still remain active. The net result is that attractive goals trigger dopamine release to many parts of the striatum at once—the accumbens, still waiting for its rewards, and the dorsal striatum, which doesn't care about rewards and just wants to act, no matter what.

What a strange collusion between hot desire and the cold, almost calculating mechanics of stimulus-response! But that is how addiction develops—from pure desire to a chronic automaticity, accompanied by hope for what is to come. That's why impulsive individuals are more likely to develop into compulsive drug users. Whether their impulsivity was granted by genetics or by a sudden romance with feel-good chemicals, it is destined to extend to compulsivity over time. When that stage comes, the blending of ventral and dorsal activation is what allows addicts who pursue their fix robotically to remain hopeful that this time will be special—maybe as special as it felt in the early days.

Trevor Robbins and his colleagues at Cambridge have been studying the shift from impulsive to compulsive drug seeking for many years, and they are often viewed as the world experts. They see the compulsive phase as true addiction, as do many others in the field. In their view, the problem no longer resides in the perceptual system—in how attractive the drug is perceived to be—but in the action system. The problem is an action command that is difficult to turn off until it has been executed. For Johnny, this was the monster sleeping under the bed during his long flirtation with alcohol. He'd been able to control his drinking before the final six months. He liked his booze, he desired it, he often drank far more than he should have. But it was not an automatic response. He did not drink mindlessly. He drank to experience more of what he wanted. And once the drinking became automatic, there was still pleasure for a while—an hour or so. But pleasure seeking did not motivate the fourth and fifth and sixth and seventh drinks. In fact, we might imagine that desire was no longer part of the equation. Rather, he drank because his dorsal striatum was converting dopamine into a behavioural command. He drank because it was too hard to stop.

Then what was the stimulus that compelled him? It may have simply been the empty glass in his hand. Or, on first waking, the awareness of being sober or nauseated. Or, as Robbins and colleagues also suggest, it may have been a mental image of the negative consequences of not drinking: the dawning awareness of what he did not want to face, the rational considerations he needed to avoid—the shame of it. Shame had been a frequent visitor through much of Johnny's life. And now it was grossly augmented by his alcoholism. Shame is one of the most painful emotions. The urge to ease it can be overwhelming, and here the self-medication model of addiction rings true. Yet we can still see desire at work—the desire to avoid pain, the desire for relief. And what about anxiety? Every morning of his self-imprisonment, Johnny was assaulted by the immediate anticipation of nausea—and of anxiety itself. Anxiety about anxiety. By the age of sixty-three, Johnny was painfully anxious about being awake from the moment he first woke up.

In OCD, often considered the first cousin of addiction, rituals are performed to avoid or to ease anxiety. There is no reward in washing your hands for the fiftieth time, or checking the light switch yet again. Compulsive behaviour is a highly automatic response to a stimulus, but it can combine with the motivational flare of the accumbens and the fear networks of the amygdala, laid down over years of personality development. The dorsal striatum does not act in isolation, and the accumbens and amygdala are its most vociferous neighbours. According to Robbins, this combination of neural forces may best explain full-blown addiction.

But doesn't this kind of neural restructuring qualify as a disease? Once a person has reached this state, the brain is no longer functioning as it did. According to Barry Everitt and Trevor Robbins, "There is nothing aberrant or unusual about devolving behavioural control to a dorsal striatal S-R habit mechanism." They remind us that this occurs in many aspects of our lives, including eating and other normal activities. When that forkful of pasta plunges toward your mouth, is it because you are anticipating its lovely flavour, or is it because that's where the fork is supposed to go? Or both? "Automatisation of behaviour frees up cognitive processes," they continue.[1] And when you think about it, that's pretty obvious. We need habits in order to free our minds for other things, like think-ing, talking, exploring, evaluating, and planning. As described in Chapter Two, self-organizing brain changes naturally stabilize into habits, and whether they are good or bad habits is a social ques-tion, not a neural one. Yet Johnny's compulsive drinking certainly does not appear normal, or healthy, or natural. How do we recon-cile a natural process of neural change and stabilization with the grotesque pattern of consumption that so often rules in addiction?

Not very easily. Which is why addiction is so troubling, so mys-terious. Yet calling it a "disease" is just too simple. And it doesn't actually explain very much. So let's focus instead on the issue of control. When habits begin to threaten our security, there is a cog-nitive override designed to keep them in check. We routinely rely on self-control to steer us away from our most precarious attrac-tions. And, indeed, there are several brain regions, in and around the prefrontal cortex, whose job specs are exactly that. They often do their jobs well, as when you tell yourself to put your fork down because you've had enough pasta. Those resources help explain why most people either don't get addicted to harmful substances or eventually pull themselves out of addiction. But self-control wasn't working for Johnny.

Why not?

Because when habits get ingrained, some of the two-way traf-fic between the striatum and the prefrontal cortex starts to thin out. Especially on roads leading to a region called the dorsolateral prefrontal cortex, which is critical for reasoning, remembering, planning, and self-control. We can think of the dorsolateral PFC as the bridge of the ship, because it steers thought and behaviour deliberately, consciously, and often skilfully, overcoming challenges and setbacks. This region becomes hyperactivated in the early stages of addiction, perhaps when people try to control or maintain the enchantment of this new experience. But then something starts to give way. We've been taught to think "use it or lose it" when imagining brain function. But in the case of the dorsolateral PFC and its role in addiction, you both use it and lose it. Over time, the dorsolateral PFC and other prefrontal control centres start to disengage from the striatum when the addictive substance is at hand. *I just feel like doing it, so why not?* Disengagement leads to disuse, and disuse leads to dissolution. In long-term addiction, some cognitive control regions may actually lose a fair number of synapses— they may become pruned, and show up on brain scans with a "loss of grey matter volume" (grey matter being the stuff of neurons and their synapses). Whether the addiction is to alcohol, meth, coke, tobacco, or heroin, grey matter volume in some prefrontal areas has been thought to decrease by as much as 20 percent. And the degree of loss appears to correspond with the length and severity of the addiction.[2] According to this view, the communication between prefrontal control and striatal compulsion isn't only constricted; it's become fragmented or inaccessible. That's not hard to imagine in Johnny's case.

Yet despite the depletion of top-down control, addiction remains very human. And very common. Compulsive habits overtake substance use, sexual excess, and gambling—all the usual suspects. But more familiar compulsive habits include nail biting and nose picking—habits that many children, even adults, find difficult to break. And spending money, watching TV for hours, playing video games all night, excessive dieting, removing unwanted hairs from various body parts, and pursuing nice-looking sexual partners who you know are going to end up causing you more grief than gratification. And becoming dependent on someone who can't be trusted—a compulsion that, in its more extreme forms, brings abused women back to their abusers and keeps men and women longing for lovers who don't love them back. Jealousy, even in relatively healthy relationships, often has compulsive qualities and remains difficult to control—sometimes impossible to control. And so, of course, does blame, the background music of so many normal marriages. And eating: that most basic of activities. Given current obesity statistics, it's not surprising that the psychiatric accountants who update the DSM are filing overeating under "addiction."

Yet few would be tempted to refer to these habits as disease.

~

When Johnny saw the imprint of his face on the kitchen floor, his personal Shroud of Turin, he knew his future was collapsing around him. The business was falling apart, his family was a distant dream, and his health was disintegrating. Even while drunk he could no longer bear to think about his failures. Not only was the drinking compulsive, but so was the psychological act of denial. He would hide the empty bottles toward the end of a session, so that he would not know how much he had drunk the next time he woke up. But where to put them? He would wrap them up in newspaper, so they wouldn't clink together, and collect them in large plastic garbage bags. Even at this point, especially at this point, he was concerned about the neighbours. He couldn't face the thought that they'd find out how far he'd fallen.

His life had become hellish. He was desperate to stop.

He found the phone number of an expensive detox centre on the outskirts of Manchester. One bleary afternoon, he called. He brought himself to an appointment. The intake worker said that he'd be dead in three weeks if he didn't quit. He wasn't too surprised.

"But keep on drinking for a few more days," she continued. "We don't want you to have a seizure. Don't come in falling-down drunk either. We'll have a bed for you next Wednesday." Wednesday turned out to be Friday. He arrived at 9:15 a.m. He has not had a drink since then—five years as of this writing.

"I was a zombie for a week," Johnny told me. He was on heavy doses of diazepam the whole week, to suppress seizures and give his body a chance to adjust gradually. He wasn't sure what other drugs they gave him. But he started to feel better. Although his hands shook for a while longer, he was now able to sleep more normally. He was able to walk without tilting. After the first week, he began to participate in the program: group activities, meetings, diaries, more meetings, and a lot of social support. Johnny stayed at the centre for twenty-eight days, as requested. I asked him about cravings during this period, and he said there weren't any. With everyone's attention focused on support and sobriety, and with the medication they continued to administer each day, the cravings had no point of entry. The program took you right out of your life, he recounted. You were in a different world: segregated, protected, and soothed.

Then he found himself back home. It's common knowledge that the period following residential treatment is when addicts are most vulnerable to relapse. For obvious reasons. But Johnny applied the determination he'd prided himself on since his youth: a determination to succeed. At first, he gave himself a daily mandate—"Just stay sober for today"—reflecting the AA slogan, "One day at a time." And things were pretty easy for the first week. Then temptation started creeping in. "You suddenly realize you're a father or husband, an employee or an employer. You remember you have responsibilities. You think: Was I a good father? Was I a good employer?" And a familiar answer followed: No, you were not. Ruminations, doubts, and anxieties filled the empty spaces, much as they had before treatment. So Johnny fought back. He used prayer, asking God to help him to get through the day sober. And he went to AA meetings, as instructed by the treatment staff. They told him he must attend fifty meetings in fifty days. He made it to thirty and then stopped going. It was exhausting, he said. "AA is brilliant for some people. But it wasn't for me."

Now on his own again, he decided that he could never stay sober until he found the reason he'd become an alcoholic. So he went to see a psychiatrist, who essentially agreed. "You're suppressing something that you won't come to terms with," he was told. So he worked to unlock the memories and feelings he'd hidden. Several months later, the psychiatrist praised him, said he was learning to analyze himself and didn't need to keep coming. But he recommended weekly sessions with a counsellor. Johnny got the name of one from a friend, then paid a handsome fee to have her come to his apartment for sessions. "She came to my place every single Thursday, for a year," he declared. "It took me three months to open up to her, and after that I wouldn't shut up." She was very skilful, in Johnny's opinion. "She had me super-analyzing myself. What did you used to do when you decided to get drunk? I sat in a corner, feeling scared." She coached and coaxed him to come to terms with the deepest wounds of his childhood: the devastating anxiety he'd endured at boarding school.

How often do we hear of the soul-destroying power of child sexual abuse? We no longer doubt its reality or the carnage it creates. But when Johnny entered his teens, over fifty years ago, it was a different world. He felt he couldn't tell his parents what was going on at his boarding school. "So I chose to be in denial," he now recognizes. "It never came to me consciously, but I used to have nightmares about it." And the result? Throughout his adulthood, whenever he'd try to get close to others, "I could never give of myself. Because I was always holding something back." Through the hard work of self-examination, he came to understand why he had always felt alone and scared. And why he'd become addicted to alcohol's promise of peace.

Since the end of his year of counselling, Johnny has continued to help himself stay ahead of the anxieties he may never completely escape. He has practiced yoga and Reiki, tried various types of massage, "anything to make me feel calmer. Anything that would bring on the feel-good factor."

~

So what made the difference? What changed his life? Was it the treatment centre he attended? Was the program the wedge that split his life into before and after? No, Johnny doesn't see the program as anything special. There was no one-to-one counselling, a lot of medication, and a treadmill of warnings and slogans, repeated day after day. He's also done his own accounting. Johnny was one of eight men admitted to the program on that particular Friday. He got to know the other seven fairly well and kept track of them as best he could. After six months, he was the only one who hadn't gone back to drinking. Three have now died. He's lost track of the other four, but he figures that's not a good sign.

Then was it the intensive psychotherapy he received, following a psychiatric dig of archaeological proportions? That certainly helped him get to the bottom of things, but was that the magic wand? Or would he have quit with any sort of support, or perhaps with none, given his degree of desperation?

Those who see addiction as a choice might argue that alcohol was causing Johnny more harm than good, that drinking gave rise to more suffering than pleasure, so he *chose* to stop. Yet it had been obvious to Johnny for a very long time that alcohol was damaging him: his social life, his health, even his sanity. Why didn't he choose to quit before? If choice were a concrete variable in the equation of

Johnny's drinking, we'd have predicted that he'd quit much sooner. The trouble with the "choice" argument is that choice isn't simple. It can't be. It arises (like all mental activities) in the slushy insides of our brains, where long arrays of habits overlap and intermingle, where associations, feelings, context, and "now appeal" highlight different goals from moment to moment, where "the right thing to do" isn't always obvious.

In Johnny's view, what most helped him stay sober was the work he did himself: the work of discovering himself, acknowledging the source of his anxieties, reinterpreting, even reinventing himself. And, maybe just a little, learning to be kind to himself. It was cognitive work and emotional work; it took effort and determination. What alcoholics need, he believes, is to become aware of their personal baggage. "They need to know why they drink, and what triggers them to drink. Otherwise," he declared in our last interview, "treatment is useless." Yet just being with other people in the treatment centre did something important for Johnny. Habits thrive on consistency—consistency in one's environment, one's daily routines, and one's social world or lack thereof. His time in treatment interrupted that regularity and offered the promise of new and different habits. And there is little doubt that Johnny's counsellor helped him to continue the courageous project of drilling into the bedrock of his past. And extracting ore from whatever he found there. A number of pieces came together for Johnny. But what brought them home and arranged them was his own will, his own intention, and the day-to-day insights that came with it. So that the interlocking habits of alcohol, anxiety, shame, and solitude could be kept apart for good.

He still misses drinking sometimes. Occasionally he feels resentful: "Why am I the only one who can't ever drink again?" But that doesn't last long these days. It ends when he tells himself, "Don't be stupid. Alcohol is like a poison for you."

∾

Compulsion may be the final stage of addiction, but compulsions are not carved in stone. Brain flesh remains mutable throughout our lives. The regions of the prefrontal cortex that can overcome ingrained habits may lose their efficacy in addiction; they may even be thinned out over time. But neuroscientists have no idea what this thinning actually means. A vast number of synapses (up to 50 percent in some regions) get pruned with normal maturation, from infancy to adulthood, allowing greater efficiency in overall signal processing. Neuroplasticity is not a one-way street, and sectors of depleted synapses make room for new synapses to grow. Importantly, that doesn't just mean *regrow*.

A study published in *PLOS ONE,* the journal of the Public Library of Science, in 2013 showed that the reduction of grey matter volume in specific regions of the prefrontal cortex, thought to progress with the length of addiction, reversed over several months of abstinence.[3] Grey matter volume returned to a normal baseline level within six months to a year of abstinence (from heroin, cocaine, and alcohol), and similar results have been found by other studies as well. But then—and here comes the first surprise—grey matter volume (synaptic density) in these regions continued to increase, beyond the normal baseline level, the level recorded for people who've never been addicted. Which probably means that top-down cognitive control regions—what I call the bridge of the ship—became more elaborate, or sophisticated, or flexible, or resilient, than those of people who had never taken drugs. It makes sense. Abstinence requires sustained and seasoned cognitive effort, and that effort grows synapses as surely as any other motivated activity. The second surprise is that, when you look closely at the brain scan images in the article, the regions of increased growth don't correspond exactly with the regions initially depleted. That could mean that recovering addicts don't just regain their lost self-control; they actually develop entirely novel strategies for self-control, based on newly acquired neural terrain.[4] Their abstinence may endure because they've become experts at self-regulation—the very skill they so blatantly lacked. This newly honed capacity may derive from teeming clusters of synapses that appear in the most sophisticated cortical regions (e.g., the lateral prefrontal cortex and anterior cingulate cortex), areas responsible for self-monitoring, self-regulation, behavioural control, and choice—yes, choice.

Addiction, like everything else we think and do, has a biological basis. And the final stages of addiction, when impulsivity gives way to compulsivity, result from synaptic patterns that reinforce themselves over countless repeated occasions—a creeping vine that eventually strangles the other flowers in the garden. Motivated repetition results in compulsion through biological changes in a brain that digs its own ruts. But that's not a disease, and it's not permanent. The research I just summarized shows that choice is biological too. And choice, while it may look "rational" upon quick inspection, is driven more by motivation than by abstract reasoning. People choose to stop when they have suffered more than enough. And when circumstances lend a hand. And when the possibility of self-control becomes as attractive—more attractive—than any other possibility, including temporary relief.

In fact, real choice is not a one-shot deal. It is not a moment in time or a fork in the road. Rather, self-control thrives, as does addiction, when new mental habits are fashioned, and rehearsed, and strengthened by ongoing self-reinforcement. Choice may indeed be an antidote to compulsion, but it is also an evolving skill, fuelled by desire—in Johnny's case, desire for the strength and freedom he had always coveted.

SEVEN

Nothing for Alice

The Double-Edged Sword of Self-Control

Alice had found herself ugly since she was four or five years old. It was just a fact she had to live with. Why else would Laurie send her away, time after time, refusing to be her best friend or any friend at all? She'd have done anything for Laurie. When she was allowed into the *Little Mermaid* game, she never asked to be a mermaid, and Laurie made it quite clear that she was not mermaid material. Laurie would always play Ariel, the beautiful mermaid who wanted to become a human so she could marry Prince Eric. Alice was relegated to playing Flounder or Sebastian, bit players in Laurie's drama.

One of the most insidious features of this friendship was its longevity. In kindergarten, Laurie would say, "I don't need you today," and Alice would go into a corner and cry. She never fought back. She didn't know how. Alice's mother made sure the two girls were placed in different classes in first grade, but they continued to see each other in Girl Scouts. Alice kept up her efforts at friendship, but she couldn't win. On one of the blissful occasions that Laurie invited her over, she made it clear that she'd called fifteen other friends first.

Alice was not even a second-string player. Then in fifth grade they were back in the same class, and Laurie's brand of contempt caught on with her inner circle, as often happens in the preteen years. Finally, in sixth grade, Alice received what she calls a hate letter from Laurie. It stated that Alice was to avoid contact with her from that time on. She was not to call her or approach her in any way. Alice was *not* her friend.

Relations with other peers were not always happier. Alice remembers playing spin-the-bottle in her early teenage years. She remembers the boys sniggering, whispering that they'd die if they ended up with Alice. Because she was so ugly. My guess is that she just looked insecure.

Maybe that's all it takes. It's enough to give off the scent of anxiety, which socially skilled teens pick up like circling wolves. Alice says her own mother was a very anxious person, constantly obsessed with her hair, which was never quite right. And with her stomach. And with her breasts, which were too small, in her mom's opinion. With such small breasts, people would notice her stomach sticking out. It was an agonizing fact she shared with her daughter often. We can't know how her mother's anxiety was transmitted to Alice. Genetics? The ominous atmosphere in her home? Witnessing how hard it was for her mother to make simple decisions, make friends, or receive scraps of affection from her husband? Alice's father was emotionally distant, as was Donna's. But while Donna's father was suicidal, Alice's father was made of steel. He was the plant manager at a textile firm and a devoted golfer. Alice's mother was simply not a part of his life. Seeing this with increasing clarity as she grew up, Alice learned to hate her father and feel sorry for her mother. She was determined never to end up like her mother: an anxious, depressed housewife obsessed with her hair and the size of her breasts.

Yet Alice couldn't ignore her own appearance, even as a child. She tried putting tennis balls in her T-shirt, hoping fervently that her breasts would one day be big enough to overshadow the flaws she knew she had. Such as her stomach, which always seemed to stick out a little. She was obsessed with fitness videos and *Sports Illustrated* ads featuring women in bathing suits. What was their secret? How was it possible to be that attractive, that strong?

Alice told me in one of our interviews that she'd been naturally slender since she was tiny, and she became increasingly slender through her high school years, a result of constant dieting. She wanted her stomach to be "toned and flat," and she wore baggy clothes to hide the minimal curvature that wouldn't go away. Still, she acquired a group of friends who were decent to her, and with whom she felt comfortable, if not completely at ease. That was good enough. The next four years were spent at a small college in the southeastern United States, where she brought her excellent mind to bear on her studies in biology and psychology. But where food was concerned, she remained helpless, bouncing from one diet to another in an effort to make her flat stomach flatter. By the end of her undergraduate years, she accepted her fate. There was nothing she would ever be able to do about her stomach.

And then she stumbled on the secret: a way to reduce the curve of her stomach to a flat plane. The summer before she entered graduate school, she became a vegetarian and got rid of ten more pounds. That brought her down to 106—and she crossed the line into anorexia.

She felt grateful, uplifted, victorious, safe, in a way she still finds difficult to explain. She felt that she had finally achieved control, a sense of purity and emptiness that was profoundly satisfying. That feeling did not subside, even when she discovered handfuls of hair in her hairbrush, hair on the floor, and, in the mirror, the reflection of someone who didn't look much like her.

~

In one way, anorexia seems the opposite of addiction. We characterize addiction by the loss of self-control, but anorexia epitomizes overcontrol. In fact, research reveals that, compared with others, anorexics are faster and better at decision making. Alice was certainly organized and strategic by nature. But the *way* she organized her life was by restricting and restraining, tightening and tightening and tightening some cog in the machinery of her inner world. How else could she have given up so much—given up mermaid status to be Laurie's slave, and held on to that relationship for so many years, despite the meagre scraps of well-being she got from it? Making herself smaller to accommodate Laurie's greed. Addicts are notorious for shucking off self-control, for abandoning it like a worn-out car. Anorexics seem to have the opposite problem: they can't take their hands off the steering wheel.

Yet anorexia and addiction are as similar as twin sisters. Both get their enormous power from compulsivity, and that compulsivity remains hidden from oneself long after it becomes obvious to others. When impulsivity grows into compulsivity, as it did in each of the accounts I've included in this book, the dorsal striatum rises from its slumber, stretching, impervious to other agendas, to insist: You must do this—now. And the self-control networks of the prefrontal cortex lose their efficiency (and perhaps eventually their synapses) when attempting to regulate it. Neither anorexics nor drug addicts exercise top-down control when it comes to the behaviour that harms them. Alice had little control over her need to starve herself. Even the clumps of hair, omens on the bathroom floor, weren't drastic enough to change her course. Nor did she feel there was anything that *needed* changing. At this time in her life, she did not feel she had a serious problem. But the acute irony of anorexia remains: the thing she could not control was her conviction that she was finally *in* control.

Anorexia and addiction are similar in another way. In both cases, it is the surrender to self-deprivation that gives rise, over the years of childhood and adolescence, to the spiralling growth of overwhelming need. Natalie deprived herself of spontaneity to avoid her stepfather, choosing depression, burying herself in her bedroom. Donna kept her mouth shut, complied with the demand to mute her impulses, in an effort to be the good girl her mother insisted on. In each addiction narrative, we can see an underlying current of anxiety, anger, or fear of rejection, and an unconscious collusion between the child and his or her caregivers to accept the burden of inadequacy. That collusion is exhausting. It creates a kind of emotional starvation.

With most addictions, this starvation rebounds into desperate and repeated attempts at self-fulfillment. The highs we get from drugs, booze, and porn are extreme antidotes to inner emptiness. Like starving rats with soaring levels of dopamine, addicts are ruled by desperation to fill the void that has grown within them. Their dopamine levels spike at the whisper of possible fulfillment. But anorexics are different, because the satisfaction of their need consists of ongoing deprivation. Anorexia, then, is an exquisite extension of self-denial, satisfying because it completes the search for perfection that began years before, dangerous because that search leads to the excavation of one's own flesh.

In another sense, anorexics are classic addicts because they relentlessly pursue a symbol. Symbols gather our cognition, our thoughts and associations, into coherent emblems, full of meaning yet consisting of very little in themselves. Symbols always represent something else. Symbols include beautiful women, flashy cars, fatherly love, financial security, even the idea of youth. Each of these is an arrow on a map, or an array of arrows aligned with each other, a direction to pursue. Each shrinks a cluster of related goals into a single goal we can chase after unambiguously. For Alice, that goal may have been attractiveness, leading to positive regard, at least in childhood and adolescence. By now it was simply self-mastery, that crystalline sense of control—a symbol far more refined, more idealized, than anything she had previously sought.

Drugs are also symbolic, in a way that is rarely if ever discussed by the experts. But what they symbolize is likely to evolve and mutate. Symbols develop, and development takes time. As we saw with Brian, the careening arousal of a meth high took many months to mean what it eventually came to mean: clarity, power, and self-confidence. So it is with the self-starvation of anorexia. What may start off as the pursuit of a particular body shape can end up having little to do with one's appearance and everything to do with restriction and self-control.

There is no single brain region or system where symbols are created and activated. But the prefrontal cortex (PFC) is a critical player, connecting present experience to the cluster of elements the symbol stands for. The part of the PFC that pulls information together—the dorsolateral PFC—is known to become highly activated when addicts are exposed to their drug of choice and when anorexics and bulimics are exposed to food (or food cues). At least until it starts to disengage, as described in Chapter Six. The perceptual cortex is also part of the team, providing the sights and sounds that dress the symbol in its garments, and so is the amygdala, where the flame of emotional meaning is suddenly ignited. Symbols pack a lot of meaning, but they also come to a point, a sensory image or a word spoken or remembered. Words and images are cues, and cues are the keys that open the locked gates of addiction and anorexia both. Cues trigger craving. Cues trigger relapse. Cues release dopamine in the accumbens, cranking the engine of desire.

That's how symbols govern our actions, from humanity's greatest accomplishments—scientific discoveries, symphonies, works of art—to its greatest tragedies—crime, persecution, and, yes, addiction. But the symbol and the passion don't always converge instantaneously. They can arise in parallel, then connect, then develop some more, as did Alice's anorexia, from a child's pride at self-control to the austere habits that defined her body and mind twenty years later. Both anorexia and addiction rely on neural networks that *grow*.

~

Whatever else it is, anorexia is a literal form of starvation. By the time Alice got to graduate school in Toronto, she was hungry at every level. So, like a majority of people with eating disorders, her "diagnosis" shifted from one fuzzy category to another. She found herself at a party early in her first year, feeling shy and anxious, not knowing anyone there. These were familiar feelings. What was new was that she was starving. She recalls standing by the food table, sort of planting herself there, taking refuge. And then helping herself to a large, rich cookie. She showed me the size of this cookie with her fingers on a Skype call. One would be plenty for most people.

That night Alice ate fourteen of those cookies, one after another. She remembers the number exactly, because she herself was astonished. "It felt so good," she told me. "It was absolutely what I wanted, what I needed." An empty place finally filled.

Binge eating soon became a regular event. Two or three times a week, she would start eating and not stop until she felt ill. What she ate was rich, or sweet, or salty, or some combination of these. It was food with impact, including huge quantities of guacamole, peanut butter out of the jar, and chocolate. Then she would take laxatives to get rid of the extra mass. She was still dieting. She was still trying to keep her weight down. But her self-control was now strained and leaking. She was stretched to the breaking point by forces pulling in opposite directions. She would be labelled a bulimic at this stage. In her mind, she was just a mess.

Binge eating resembles substance addiction more precisely than does anorexia. Binges are compulsive acts of input, filling the self with something badly wanted, desperately needed. The dorsal striatum—which induces behaviour automatically, mindlessly—is turned on in anorexia and bulimia just as it is in compulsive substance use. But what about the dorsolateral PFC, the bridge of the ship, which goes into action when booze and drugs are spotted—at first—and then powers down, becoming relatively dormant, as the habit coalesces? According to Nora Volkow, a leader in the neuroscience of addiction, the dorsolateral PFC disengages from the striatum in obesity and binge eating, just as it does in drug addiction. So there would be a breakdown in communication between the part of the brain that pursues goals and the part of the brain that could control that pursuit. Is that what causes the loss of synapses in some prefrontal regions, as addictions and eating disorders develop? We don't know yet. But Volkow is convinced that the lack of control common to both addiction and eating disorders results from a breakdown in communication between prefrontal and striatal structures.

Like many addictive behaviours, Alice's binges were triggered by periods of intolerable anxiety. Graduate school was a lot more difficult than she'd expected; she was afraid of failure, and she was lonely. She now attended a fine department at a large, urban university, but she had no friends there. Her only friend was the man she had just married, and that made things even more complicated. The bingeing disgusted her, and she knew it disgusted him. She did not want him to see what she'd become. She watched herself in amazement as she gorged on food she normally wouldn't touch. Like the alcoholic or crack addict, she felt she was a slave to desires more powerful than she was. She felt she had to do this. And if he tried to talk her out of it once she'd already begun, she would slap him away, then dig her fingernails into her own skin and lie on the floor, chanting, "I hate myself, I hate myself." Once she'd begun to binge, she only wanted him to leave so that she could continue. By herself. Get it done.

But it wasn't just food that attracted and repelled her. It was the sequence of intricate steps she followed more and more faithfully each time she entered a binge phase: her ritual. The striatum is ignited by cues: stimuli (words and images) or events that foretell the addictive act. And symbols grow like glistening spiderwebs, each with a central hub connected to multiple anchor points. Rituals combine actions and goals into strings of events that are themselves symbols. And they are powerful.

OCD is often considered a pure form of the compulsive drive that marks the final stage of addiction. And we know that people with OCD arrange their lives into repetitive rituals. So it shouldn't be surprising that most addicts find comfort, but also excitement, in the rituals that lead up to the act itself. The fastidious preparation of the heroin injection; the arrangement of a glass surface and a blade to pulverize crystals of cocaine; the dressing up that precedes a night of gambling or sexual excess. The rituals become gratifying in themselves, part of the webbing of symbolic significance.

Addiction researchers like Kent Berridge find that cues take control of the dopamine pump with astounding efficiency. The dopamine baton is passed back from one cue to the next, to the next, to the next, each one predicting the one that predicts the eventual moment of release. The dopamine wave that once crested just before the addictive act gets retracted, step by step, so that otherwise neutral events become laced with meaning, resolve, and excitement. For Alice, as for others with eating disorders, counting calories was the prized ritual. It packed a tremendous amount of significance into a series of trivial acts: toting up numbers that might allow for a midday bowl of lettuce or, if she was very good, a feast later on. If she was still at 400 calories by dinnertime, she could indulge in a burrito without guilt—a well-earned reward. So she counted calories with every bite, making lists, adding and subtracting values. She checked the bathroom scale several times a day. There were so many spaces to fill, so many particles to keep track of. It was good to be in control again. Until the next binge, when all control would vanish.

For several months, her eating habits rode on a merry-go-round of anxieties that would not slow down: performance anxiety clawing at her before a conference presentation, social anxiety when she found herself among unfamiliar people, and the agonizing mix of the two—trying to mind her social behaviour, to keep up appearances, in front of family at Christmastime, in front of her fellow students, at a lab meeting with her supervisor. These were times when her vulnerability began to explode in slow motion. She would try to hold it back for a few hours or even a few days. But then her resolve disintegrated and the compulsion broke through, carrying her like a matchstick in its swell. She always felt ashamed when she broke down, even before she broke down, knowing it was imminent. Her shame grew in strength and ferocity, battering the walls of its locked cell until it ran amuck. A monster familiar to most addicts. Alice could handle her shame. She had known it in one guise or another throughout her life. Yet it was grotesquely magnified when her husband caught her in the act. She knew how revolted he must be, just watching her. And before long he let her know that their marriage was in peril.

~

Alice's pattern of sustained self-control, followed by the sputtering loss of control, is typical of many addictions. And it's been given a name by psychologists who have studied it in the lab for almost two decades: ego fatigue, or ego depletion. That sounds old-fashioned, even Freudian, and contemporary researchers sometimes rename it self-control depletion, which is exactly what it is. But psychologists have had a hard time figuring out how it works. Some people can maintain some kinds of self-control for indefinite periods. And yet, especially in tasks that require you to suppress your desires, hide your emotions, or ignore important information, self-control begins to blink and fizzle like a dying light bulb.

Roy Baumeister collected the earliest evidence of ego fatigue in a study published in 1998. All participants came to the lab hungry, and all were seated in front of a bowl of fresh cookies and a bowl of radishes. Half were told to eat as many cookies as they wanted, and the other half were told to eat as many radishes as they wanted—but no cookies. A few minutes later, this second group could not perform as well on tasks that required self-control and sustained effort. And they often gave up sooner. The same effect was found for participants asked to avoid emotional expressions while they watched an emotionally compelling video. Afterward, cognitive control was less available. This type of task has been repeated in various forms hundreds of times, and the results are indisputable: after suppressing one's desires, impulses, or habitual responses for some period of time, the cognitive machinery of self-control gets worn down and dysfunctional.

But why? We used to believe that ego fatigue was caused by the exhaustion of the chemical resources needed for brainpower. But the data don't support that view. Contemporary researchers, like Michael Inzlicht at the University of Toronto, propose that people stop trying because frustration builds up and letting go simply becomes more attractive. But that's still not much of an explanation. Then along comes neuroscience. As Inzlicht himself discovered, brainwaves that indicate self-regulation, measured while people were trying to avoid mistakes, became weaker with ego fatigue, but only when they tried to *suppress* their feelings. Other kinds of emotional control, like *reinterpreting* emotional events, did not cause ego fatigue. This kind of control, which requires shifting one's perspective on whatever seems challenging, did not hamper performance or disturb brainwaves. When Alice's physical and psychological hunger threatened to burst through her dieting, all she could do was suppress her desires. There was no way she could find to reinterpret them.

Neural experiments using fMRI brain scans give us a more precise picture. Ego fatigue increases activation in brain areas devoted to emotional meaning—the amygdala and orbitofrontal cortex—while simultaneously reducing the communication between these areas and the dorsolateral PFC, where information, insight, and self-control are brought together. Recall that this disengagement of the dorsolateral PFC from the motivational core of the brain has been found repeatedly with addicts of all stripes. And Volkow and her team have found the same disconnect with binge eating. The dorsolateral PFC goes offline, self-control breaks down, and Alice switches—after two days of intensive self-restraint—from dieting to bingeing.

The argument in her head often reached a deafening pitch: Don't think about it, don't think about it . . . but it will taste so good. She would try to push it away, but intrusive thoughts pierced her reverie unmercifully. It was like trying not to think of an elephant, she said. She would try to maintain control by throwing away foods that might be included in a future binge. Everything that looked pleasurable went in the trash. Then, in her words, "it was almost like my subconscious was thinking about it. I'd sometimes pull it out of the trash. I felt really, really guilty, but the element of relief overpowered the self-control." Sometimes she'd sit and debate it, rationally, sensibly, for a few minutes, but those minutes could turn into hours. What she described to me was intense, mushrooming anxiety. It seemed she'd never be free of it. It will just keep haunting me, she told herself. It will still be there tomorrow . . . there's no way to hold it back. Sometimes the binge would start up slowly, with a relatively innocent yogurt cup. But her rituals would shift gears, and she'd find herself putting all sorts of toppings on that innocuous slab. Then came the downhill plunge into an all-out binge. She would stop at the convenience store and get all the wrong things. Then go home and eat them.

Ego fatigue is the scourge of addicts. They try—sometimes heroically—to resist their impulses to engage in their addiction. But they may have tried too hard, for too long, and in the wrong way. Suppression of impulses doesn't work, especially while those impulses are ballooning by the hour. And the Achilles' heel of ego fatigue partners up with the Achilles' heel on the other foot: now appeal (delay discounting). As self-control frays to the breaking point, it becomes terribly hard to avoid plunging into immediate rewards, as Alice did with her chocolate and peanut butter, at the expense of future rewards, like living a normal life. When self-control is unravelling, when dopamine synapses are buzzing throughout the striatum, those immediate rewards fill the radar screen, and resisting them becomes the source of a growing anxiety—anxiety that can only be stanched by repeating the forbidden act. Because you just know it's not going to get better otherwise.

~

Matters finally came to a head for Alice. One evening she found herself at a party, standing by the food table, a position she'd come to adopt automatically. She was stuffing herself on chips, cookies, cheeses, guacamole, and whatever else beckoned to her from the false security of the table. She couldn't stop, and the shame overwhelmed her. There was no excuse, no possible rationale, for this kind of behaviour. She ran out of the room, down the hall, rode down the elevator, then ran out of the building. At the first convenience store she stopped and bought a large pack of Oreos. Then back to her own apartment, where she could be alone. A kind of peace descended with the click of the lock. Now she could settle into real abandon, not the agonizing compromise she'd endured at the party. Now she started stuffing in earnest. She dipped the Oreos into peanut butter and devoured them. She found bagels, covered them with cheese, popped them into the microwave, but then opened the door before the timer went off. They're done enough to be eaten! She burned her lips and tongue, but she could not stop. In fact, the pain of the burn fit neatly into the contour of her personal hell. She deserved to suffer. Such pleasure could not exist without pain.

And then she saw herself from some other vantage point: This is terrible, this is awful. That night, standing in the harsh light of the kitchen, alone, the party left impossibly far behind, she told herself that she didn't want to do this anymore. She was in agony. Her stomach was convulsed with pain. Her tongue and lips were scorched. Still it seemed she could not get the food in fast enough. This is the last time. I have to get it out of my system, now, tonight—and have done with it. By now the pattern had become clear: every time she tried to restrict her eating too much for too long, the bingeing would follow. Ego fatigue was rotting the bridges she continued to repair. Something had to give. And so this night had to come. And it continued, until she found herself writhing on the floor, then throwing up in the toilet. Involuntarily. It had become impossible to keep eating.

Her husband came home while she was still on the kitchen floor. He cried. "I can't watch this," he said. "I can't save you. I can't be there to love you if you can't love yourself. You've got to make a decision if you want to live or not, but I can't do it for you." Then he left the apartment. Afterward, she lay down on the sofa in the

living room, as straight as possible. She could not sleep. It still hurt
too much.

$$\sim$$

The next day Alice began the complicated process of signing up
for university counselling services. She also started a search for
groups—groups for women with eating disorders. She was on automatic pilot at this point. There was no other option. She discovered an organization housed nearby, through its website. It was
only a half-hour away by bus. Luckily, an open house was scheduled for the following week. Her husband took her. She was very
nervous about sitting in a room with other women and talking
about her issues. But the anxiety began to abate as soon as she met
the receptionist, who looked at her warmly and spoke comforting
words. A week later, there she was, surrounded by women of all
ages, each with her own story to tell. And before long it was Alice's
turn to talk.

"I cried so much," she told me. "I was a mess." But even in that
first meeting, despair mixed with relief. There was a completely
novel sense of being understood. No one had understood until now.
She was able to tell her secrets. No one was shocked. But she also
cried at the irrational destruction she'd caused herself. How had it
happened? Others felt the same way: Yes, I do this terrible thing to
my body, but I don't know why.

In subsequent meetings Alice heard about an amazing variety of
eating disorders. There were so many variants, and such explosive,
unpredictable sensitivities. Participants were discouraged from
talking about the particulars of their problems, because those particulars might trigger a psychological avalanche for another group
member or a bizarre competition as to who had it worse. Details
about weight, calories, and body shape were forbidden. Everyone
seemed to be on the edge of their seats. Instead, pat formulas were

used by all: "When I engage in my eating disorder, I feel . . . " It was not a twelve-step program in any way, Alice insists. But the meetings were run according to stringent rules. And the rules helped.

Alice attended the group once a week for about a year. She engaged in personal counselling for part of that time as well. It took a while for her patterns to change. She would do well for three months, then in crept the rituals, calorie counting, and even bingeing. Never as bad as it had been before, but still . . . And then she would talk about it in group the following week. I had a slip-up, she'd say. Rather than recite the grisly details, she was encouraged to talk about being sad or scared, about social interactions, being harassed at work, school stress, marital stress. The general theme of the group soon became clear to her: "Most of us just felt inadequate and scared and just wanted to be loved."

In our final interview, Alice told me that her marriage had been improving ever since she joined the group and her eating problems began to abate. While she pulled back from her cliff edge, her husband learned to be less anxious and less vigilant about her relationship with food. "I'm not always in an eating-disordered mode," she said. As he became more trusting, she became more effective, more confident in her progress. It began to feel like she was making these changes for herself, not to assuage his feelings. "I don't want him or anyone else to focus on it. I want it to be something that I work through and achieve. Sometimes I have struggling days, and I'd rather not have someone pointing at me."

One of her greatest challenges has been dialogue with other women. And although she didn't make this connection, I can't help but think of the conversations, focused on body shape and attractiveness, she'd had with her mother as a young girl—conversations that seemed to set the stage for her growing determination to restrain and restrict. During the early phase of her recovery, her mother-in-law was the greatest challenge. "All she wanted to talk about was counting calories," Alice told me. She would tout the glories of her 1,200-calorie-per-day diet, and Alice would just cringe. Why would you want to do that? It's just unhealthy, she'd say to herself. Or she might reverse polarities and think: Why does she get to do that and I don't? Alice was ready to argue, to dispute the ridiculous norms her mother-in-law seemed to endorse. But she imagined that such arguments were somehow "anti-female." In her words, "talking with my mother-in-law was like choosing footsteps across a sea of thin ice"—but that trek may have begun in childhood.

Now, several years later, Alice rates herself as "mostly okay." "I still slip up sometimes," she says, "but I don't feel so guilty about it." She finds conversations among women to be a lot less provoking. In fact, she is working toward softening her black-and-white perspective, allowing herself to accept that women who are dieting do not necessarily have eating disorders. "My struggle," she concludes, "is to stop being afraid of food."

These days she finds that self-control is not some overarching harness she needs to apply to her impulses. She says she is more aware of what she is doing and what she is feeling from moment to moment, better able to shift her trajectory when she needs to. But her slip-ups mostly occur when she's feeling sad or lonely. That part hasn't changed.

For Alice, recovery isn't all sunshine and rainbows. "Some girls come out of treatment and say, everything is wonderful, I went kayaking this weekend, I love my body, I love myself now." That's not how it is for her. "I don't go for this love-my-body campaign," she told me. "I don't love my body, but I don't hate it as much as I used to. I don't think about it as much, and that's good."

~

Some addiction experts espouse a horse-and-rider model. They see cognitive control or self-control as the rider and the horse as one's impulses. These "dual process" models have been fashionable in psychology for several decades, and they are frequently applied to addiction. Addicts have very strong impulses, and they have lost the ability that most of us have to control them. The rider has lost her grip, the reins are slipping, and so the horse runs wild. Or else the horse is just too strong to be tamed and so the rider gives up. Notice that both of these interpretations lend themselves to the disease concept. Because, either way, a wild, riderless horse will wreak havoc until it is constrained or sedated.

But control just isn't that simple. Cognitive control is a regulating signal that emerges from many parts of the brain and fans out to many other parts. It includes different channels, playing very different programs. The dorsolateral PFC does indeed perform some high-wire acts of cognitive organization, integration, and selection. But rather than view it as the rider of a rebellious animal, we should see it as the bridge of a ship—where broad perspective allows for long-range planning as well as short-term steering. There are other prefrontal regions, one around the lower midline, and a pair of regions, one at each side, that generate more primitive kinds of control. Plus the ventral (southern) part of the anterior cingulate, just behind the prefrontal cortex, which tries to control behaviour from the midst of an emotional maelstrom. How are we to understand self-control and the loss of self-control, given these diverse systems?

Neuroscience tells us that both addiction and eating disorders show hyperactivation of the dorsolateral PFC control system, the most sophisticated of the bunch, followed by deactivation—or, more accurately, disconnection—of this system from the motivational core of the brain. Alice's compulsive bulimia, and late-stage addiction in general, would show up as reduced activity in the dorsolateral PFC and increased activity in the striatum, at least when temptation is at hand. Over time there would be less and less communication between the two structures. But that's almost precisely the picture we see with ego fatigue. Ego fatigue shows the same disconnect, between the bridge of the ship and its motivational engines, and the same loss of control in volitional behaviour. The momentary state of ego fatigue now looks like a microcosm of the enduring state we know as addiction.

Addicts' and bulimics' attempts at self-control are continuously undermined by ego fatigue. And the neuropsychological literature helps us understand why. Pure suppression doesn't work. It would seem that more primitive forms of self-control, powered by lower regions of the prefrontal cortex, have taken over from higher regions such as the dorsolateral PFC. But all *they* can do is inhibit impulses, yank those reins mindlessly when anxiety surges. Bad move. That's how you aggravate the horse.

Putting these pieces together, it all starts to make sense: the dorsolateral PFC gets disconnected and deactivated because it's been taken off the job. The powerful urges arising in the striatum seem to demand suppression, but that's not on its resumé. So the job of self-control is relegated to other, more ventral control systems. And those work less effectively. Like when you use the wrong muscles to hold your stomach in or your head up: you are cruising for a serious backache. The buildup of craving, the sense of impending failure that preceded Alice's binges, sometimes went on for days. We can imagine her battling her urges bravely every hour during this period—relying, unfortunately, on the wrong parts of her brain.

And the horse is no simple beast either. In Chapter Six I described the different levels of the striatum, stretching from pure impulse at its southern tip to pure compulsion in its northern, dorsal region. These levels usually work together (remember the image of the maypole). Although Alice's bingeing was clearly compulsive, it was also impulsive—which is to say enjoyable—at least at first. As was Johnny's drinking for an hour or so. Add to this complexity the ravages of now appeal, when the intention to act is mesmerized by the immediate, and long-term consequences lose their potency. More striatal bravado and less dorsolateral dominion, to be sure, but perhaps the horse has *become* the rider. As Everitt and Robbins remind us, "There is nothing aberrant or unusual about devolving behavioural control to a dorsal striatal . . . mechanism."[1] In fact, it's a common occurrence in everyday life, when habits—like checking the stove and turning off the lights—stop asking for permission and proceed on automatic pilot.

We might wonder if it even makes sense to view addicts as lacking control. Recall that Donna exercised meticulous control, beaming smiles and hiding her needs and resentments for most of her life—despite her relentless quest for pills. And anorexics clearly wield incredible control despite their unquenchable hunger. For Alice and all the others I've written about, powerful symbols are kept in mind and pursued relentlessly, preparations or rituals are followed with exquisite precision, and alternative goals are thrust aside by the singular intention to do the one thing that matters most.

If you still like the horse-and-rider idea, you should consider that there are many riders trading places, each with different skills; the horse may shift from a gentle friend to a raging beast from moment to moment; and it's never quite clear who is controlling whom.

∿

Yet there is something precious about the higher levels of prefrontal control, exemplified by the dorsolateral PFC—the bridge of the ship. That kind of control is conscious. It uses conscious attention to adjust one's actions in a way that will ultimately be most beneficial. It can overcome now appeal, partly by imagining a valued future and making it accessible. And it can overcome ego fatigue. As mentioned, the best way around ego fatigue is to shift perspective and reinterpret one's emotional situation. That requires moving beyond suppression to insight, which relies on software already installed in the dorsolateral PFC . . . though it may be ready for an upgrade.

Alice had to reestablish the connection between the bridge of the ship and the motivational engines of her brain—after they had been segregated, relatively speaking, for years. She had to plan her recovery. She had to find help, and she had to find the kind of help that was right for her. She had to motivate herself to get to her group every week. And she had to talk about her feelings, frightening as that was. These were effortful steps, guided by insight and a major shift in perspective. But also powered by desire. Her eating disorder had become her nemesis. What she craved now was, in her words, "being a normal person, being part of other people, being spontaneous. Holding on to the idea that my body is the last thing that people are going to focus on. And people might actually like me for me." The image of being liked, recurring unexpectedly after she'd completed group therapy, felt almost like a religious experience, she said (though she hastened to add that she is not a religious person). It came as a jolting sense that "I belong, I'm okay"—a change in perspective so absolute that we can barely imagine it. Alice had not liked herself since childhood.

What Alice had to do, and what all addicts have to do to move beyond their addiction, is reconnect desire with the higher levels of cognition. They have to reconnect their thinking and feeling. And they have to discover their own motives rather than heed the demands of real or imagined others. Often following a period of great suffering, they have to *want* to refashion their lives, and they have to want that more than another dose of the substance or behaviour they've been pursuing for so long. That's how the synapses between the striatum and the dorsolateral PFC are rekindled, extended, and strengthened. That's how grey matter volume returns to isolated control regions: not only back to normal, but beyond normal levels, as described in Chapter Six. When habits lose their strength, when synaptic traffic finds old and new routes between the striatum and the cortex, it's not that self-control suddenly appears; it's that self-control changes in character—from an imposition to a desire, from a heartless reflex to a heartfelt wish.

EIGHT

Biology, Biography, and Addiction

Connecting the biology of addiction to the experience of addiction is no simple matter. Neuroscientific concepts distil information from hundreds of thousands of experiments, each one conducted on dozens of individuals, using complicated technologies that require elaborate number crunching, involving hundreds or thousands of data points for each individual. Condensing all this into trends and averages is how the life sciences make sense of, well, life, including the subtlest changes in our most precious organ, the brain. Whereas individual experience is just that. It usually involves almost the opposite process: a lot of reflection on singular moments in a singular life, when something special or meaningful seemed to take place. To analyze experience is to apply personal insight to individual perceptions, rather than combine myriad recordings until something meaningful pops out. These are radically different kinds of data. And although there are now attempts to get individual subjects to report on their experience *while* their brain activities are being recorded, this methodology is in its infancy. There's a great deal further to go.

I've tried to take our scientific knowledge of what goes on in the brain, during the progression through addiction and recovery, and put it together with the intimate accounts shared by people who have lived through that very process. I have no doubt that the brain changes recorded by scientists are taking place at the exact same time as the experiential changes reported by flesh-and-blood people. How could it be otherwise? This book takes that correspondence seriously and attempts to build a new understanding of addiction by examining it closely.

But before clinching this new view of addiction, we need to revisit the way it's been understood up till now. The task is to overthrow myths and biases and replace them with a framework for looking at addiction more realistically, without ignoring its biology *or* its psychology. In other words, we need to be clear about what addiction is not—then concentrate on what it is.

WHY ADDICTION IS NOT A DISEASE

In its present-day form, the disease model of addiction asserts that addiction is a chronic, relapsing brain disease. This disease is evidenced by changes in the brain, especially alterations in the striatum, brought about by the repeated uptake of dopamine in response to drugs and other substances. But it's also shown by changes in the prefrontal cortex, where regions responsible for cognitive control become partially disconnected from the striatum and sometimes lose a portion of their synapses as the addiction progresses. These are big changes. They can't be brushed aside. And the disease model is the only coherent model of addiction that actually pays attention to the brain changes reported by hundreds of labs in thousands of scientific articles. It certainly explains the neurobiology of addiction better than the "choice" model and other contenders. It may also have some real clinical utility. It makes sense of the helplessness addicts feel and encourages them to expiate their guilt and shame, by validating their belief that they are unable to get better by themselves. And it seems to account for the incredible persistence of addiction, its proneness to relapse. It even demonstrates why "choice" cannot be the whole answer, because choice is governed by motivation, which is governed by dopamine, and the dopamine system is presumably diseased.

Then why should we reject the disease model?

The main reason is this: Every experience that is repeated enough times because of its motivational appeal will change the wiring of the striatum (and related regions) while adjusting the flow and uptake of dopamine. Yet we wouldn't want to call the excitement we feel when visiting Paris, meeting a lover, or cheering for our favourite team a disease. Each rewarding experience builds its own network of synapses in and around the striatum (and OFC), and those networks continue to draw dopamine from its reservoir in the midbrain. That's true of Paris, romance, football, and heroin. As we anticipate and live through these experiences, each network of synapses is strengthened and refined, so the uptake of dopamine gets more selective as rewards are identified and habits established. Prefrontal control is not usually studied when it comes to travel arrangements and football, but we know from the laboratory and from real life that attractive goals frequently override self-restraint. We know that ego fatigue and now appeal, both natural processes, reduce coordination between prefrontal control systems and the motivational core of the brain (as I've called it). So even though addictive habits can be more deeply entrenched than many other habits, there is no clear dividing line between addiction and the repeated pursuit of other attractive goals, either in experience or in brain function. London just doesn't do it for you anymore. It's got to be Paris. Good food, sex, music . . . they no longer turn your crank. But cocaine sure does.

So how do we know which urges, attractions, and desires to label "disease" and which to consider aspects of normal experience and brain change? There would have to be a line in the sand somewhere. Not just the amount of dopamine released, not just the degree of specificity in what we find rewarding, not just the (lack of) availability of top-down cognitive control: these are continuous dimensions. They don't lend themselves to two distinct categories: disease versus good health. Some authorities apply the disease label when the pursuit of a drug, drink, or activity seriously interferes with one's life. But again, where should we draw the line? The lover who turns you on may be married to someone else. And sports fans have been known to beat each other up, get arrested, and ignore their familial responsibilities when the excitement runs high. Again, "addiction" doesn't fit a unique physiological stamp, like "disease." It doesn't even fit a social category, like "fan" or "lover." It simply describes the repeated pursuit of highly attractive goals when other goals lose their appeal, plus the brain changes that condense this cycle of thought and behaviour into a well-learned habit. To put it differently, there is little benefit in calling addiction a disease, because "disease" and "normality" are vague, overlapping categories, usually delineated by psychiatrists who don't have access to individual experiences *or* individual brains.

The brain changes with all learning experiences, and it changes more rapidly and more radically in response to experiences with high motivational impact. Such changes lead to the formation of habits—neural and behavioural habits—habits that become self-perpetuating and self-stabilizing. These habits replace other, less compelling activities. That's just the way the brain is designed, by the rough tools of evolution, so that it can work as efficiently as possible. As with upright spines, it's not a perfect design and it sometimes leads to pain. (Evolution was never very concerned about suffering.) But neither backaches nor addictions are well served by the term "disease." Brain change—even more extreme brain change—does not imply that something is wrong with the brain. However, it may imply that a person has not been using his or her brain to best advantage, too often pursuing a single high-impact reward and letting other rewards fade into insignificance.

ADDICTION WITHOUT SUBSTANCES

One of the greatest blows to the current notion of addiction as a disease is the fact that behavioural addictions can be just as severe as substance addictions. The party line of NIDA, the AMA, and the American Society of Addiction Medicine remains what it has been for decades: addiction is primarily caused by substance abuse. But if that were so, why are addictions to porn, sex, Internet games, food, and gambling so ubiquitous? And why do "disorders" characterized by too much of any of the above show brain activation patterns that are nearly identical to those shown in drug addiction (e.g., overactivation of the striatum when cues are present and anticipation is high, and deactivation of some prefrontal regions when indulgence takes the lead)?

Perhaps surprisingly, the most recent rewrite of the DSM avoids the term *addiction* with respect to drug and alcohol dependencies. Rather, it calls them substance-use disorders (SUDs). But let's not mince words: substance-use disorders are precisely what most of us mean by "addiction." Instead, the current DSM refers to gambling as an addiction. Does that mean that gambling is more addictive than heroin? And "Internet gaming disorder" has been added to the list, as a "condition for further study." Which means they're not quite sure what to call it, but it seems to be in the same ballpark as crystal meth. The psychiatrists who continue refining the map of mental disorders may be almost as confused as their patients. Or, to put it more kindly, psychiatrists are becoming aware that addictive issues are defined by behavioural patterns, not particular substances. That's a step in the right direction.

People pursue certain activities repeatedly, often with little control, because those activities are highly attractive. That description can cover anything from spending sprees to helicopter parenting to jihadism, and so on and so on. But there is one very normal human endeavour that most of us recognize as the epitome of blind desire and recurrent pursuit: falling in love. Lovers think obsessively about their love object, exaggerate his or her positive qualities, and avoid thinking about future repercussions. Romantic love (but also parent-child love, and even perverse forms of love including fetishism, sadomasochism, etc.) can easily become compulsive, difficult to control, and overly focused on the immediate, with little regard for the long-range forecast. The parallels with addiction are obvious, at least in the realm of psychology and behaviour.

But a look at the neuroscience of love also reveals some remarkable similarities—this time in the realm of biology. It is generally agreed that "increased levels of central dopamine contribute to the lover's focused attention on the beloved and the lover's tendency to regard the beloved as unique."[1] In fact, several researchers have examined the love-and-addiction link directly. James Burkett and Larry Young reviewed much of this work. After emphasizing that "all known drugs of abuse cause dopamine release in the nucleus accumbens," they went on to summarize what we know about the neurophysiology of pair-bonding in mammals:

> Like with drugs of abuse, mesolimbic dopamine [which ends up in the striatum] is a major contributor to the formation of pair bonds in prairie voles and particularly in the nucleus accumbens region. Mating has been shown to cause dopamine release in the nucleus accumbens in rodents. . . . In prairie voles, pair bonding between

mating partners is prevented if dopamine receptors in the nucleus accumbens are ... blocked. ... Furthermore, nonspecific activation of dopamine receptors in the nucleus accumbens is sufficient to induce pair bonding, even if no mating occurs.[2]

The neural and behavioural parallels between love and addiction are far more complex than indicated by this brief summary. And by the way, not many species of mammals are (even slightly) monogamous, so we have to rely on small furry creatures to get our facts. But after reviewing slews of studies, Burkett and Young conclude as follows:

> In the nascent phase of addiction, large amounts of sensory information are gathered about the object of addiction. In substance addiction, this applies to the sensory modalities appropriate for the drug: the taste and smell; the particular experience unique to the drug; and the context in which the drug is taken. With partner addiction, this information is primarily social: looks, touches, words, scents, the shape of the body and face, and possibly sexual experiences. When these early interactions with the object of addiction produce rewarding outcomes, dopamine is released in the nucleus accumbens, which acts to increase the salience of incentive cues that predict the reward.[3]

Thus, love and addiction are characterized by many of the same psychological and neural features. In fact, these authors use the phrase "partner addiction," and that certainly seems accurate for many couples. Conversely, people who have struggled with substance addiction often describe their attraction to a particular substance as feeling like being in love. Brian lamented that quitting meth was like turning his back on a friend or lover. For him, the flavour of loss was the same. He'd even come to see meth as a replacement for Vera, a woman he had loved for years. Now he had to go through the heartbreak a second time.

I don't think there's much more to add to this comparison. If addiction is a disease, then so, apparently, is love.

Here are a few more reasons *not* to call addiction a disease, in summary form:

- The measurable brain changes that characterize addiction usually disappear when people stop using. (Though we can reason that more subtle brain changes remain—as they must also do for ex-lovers.) This is true of non-substance addictions as well, so brain changes in both directions have nothing to do with substances per se.

- The loss of synaptic density in certain prefrontal regions is often considered the golden proof that addiction is a disease. But all that's going on here is pruning. Synaptic pruning is one of two primary engines of normal cortical development; it is known to result from learning, and it generally increases neural efficiency.

- The environmental factors that predispose people to addiction have been well understood for decades. Yet the role of genetics continues to be emphasized by the medical community, while the role of experience (including life events and quality of life) continues to be played down.

- Psychological and environmental predictors have a lot more to do with how we *experience* our environments than with the actual nature of those environments. But diseases are based on exposure, not experience.

- The powerful attraction to addictive drugs and activities is a response to some degree of psychological suffering, including social isolation and recurring negative emotions. The "Rat Park" studies show that even rats will voluntarily withdraw from narcotics when their environments become more livable, as did most Vietnam vets when they got back from the war.

- While pharmaceutical medications can ease withdrawal symptoms, which are short-term for some drugs and nonexistent for others, they can only reduce addictive urges while still in the blood. They suppress urges by replacing addictive chemicals with other, closely related chemicals that can also be addictive.

- An exception to the above may be drugs that disrupt the dopamine system. But blocking dopamine metabolism also curtails emotional feeling, goal pursuit, libido, and overall drive. That's the unfortunate side effect of many antipsychotic drugs.

- In fact, treatment centres that refer to addiction as a disease use medications almost exclusively to ease withdrawal symptoms. Then they generally switch over to twelve-step methods, which have nothing to do with medicine.

- Many of those who treat addiction believe that the most effective tools target cognitive and motivational processes such as self-determination, insight, willpower, and self-forgiveness. There is no disease that can be arrested by tapping such processes.

- Social processes such as support and love, and contemplative processes such as mindfulness meditation, have also been shown to be effective. Again, we can't cure diseases through such processes (as far as we know).

IF IT'S NOT A DISEASE, THEN WHAT IS IT?

In Chapter Two I drew a pretty extensive picture of how brains change as people (and their habits, their personalities) develop. The repetition of particular experiences modifies synaptic networks. This creates a feedback cycle between experience and brain change, each one shaping the other. New patterns of synaptic connections perpetuate themselves like the ruts carved by rainwater in the garden. The take-home message? Brain changes naturally settle into brain habits—which lock in mental habits. And the experiences that get repeated most often, most reliably, and that actually *change* synapses rather than just passing through town, are those that are most compelling. In fact, desire is evolution's agent for getting us to pursue goals repeatedly. So intense and recurrent desires will naturally change the *rate* of learning. They will speed up the feedback cycle between experience and brain change.

This *accelerated learning* is bound to create entrenched habits, like those comprising addiction, for several reasons:

- Desire focuses attention, primarily through the activation of circuits between the striatum and other brain parts. Desire + attention = attraction; attraction propels engagement and repetition—and therefore learning.
- A feedback loop that cycles faster will generate more of whatever it generates, thus accelerating its rate further. This is the well-known snowball effect.
- Any growth process that is speeded up enough will outrun its competitors. Since synapses fade and disappear when they are no longer in use, and since addictions are pursued at the expense of other goals, addictive habits come to usurp habits incongruent with addiction—like generosity, integrity, and empathy.
- Addiction offers highly desired rewards that soon disappear, thus priming the desire pump repeatedly.

This last point deserves close scrutiny. We know that synaptic patterns get reinforced with each repetition of the same kind of experience, whether it's playing the piano, baking bread, or smoking crack. And we know that repetition boosted by motivation is the strongest driver of synaptic shaping. Every time desire initiates another run for drugs, drink, porn, or gambling, it refines the network of synapses that anchor the addiction. So imagine the potency of a longed-for reward that only lasts a few hours. In its wake it leaves loss, disappointment, and often depression. Then desire naturally flares again, in the form of longing or craving, and the cycle is very likely to repeat itself.

In this way, addictions are based on false advertising. The striatum responds with eager anticipation to the glitter bestowed by our wishes and fantasies. But the high is never as good as promised and, worse, it doesn't last. For Brian, it was sparkling crystals of methamphetamine in a brimming pipe—tipping toward exhaustion. For Johnny, a glass of rum crackling around ice cubes—then the fade-out to unconsciousness. Alice was greeted by intense physical discomfort and searing shame after every binge. With these and other addictions, enjoyment fades quickly as drugs wear off, drinking sedates, the money's spent, or sexual pyrotechnics become boring. And then . . . there it is again, calling to us, a few hours later or the following day. Fragments of reward are tasted repeatedly, followed each time by loss and renewed craving. With the flame of desire rekindled so often, the same neural passages get dredged again and again. The result is accelerated learning.

But the onset of addiction doesn't always look like an accelerated learning curve. The biographies of Brian and Donna show us that people can take drugs for a long time before they become addicted. Brian's use of stimulants and Donna's use of opiates had little motivational thrust for that initial period. Brian was trying to remain awake and alert so he could accomplish work goals, and Donna was taking Vicodin for pain long before she took it for pleasure. Nevertheless, both Brian and Donna reached a turning point at which the learning curve must have risen steeply. For Donna, this occurred more than once, since there were periods of calm between

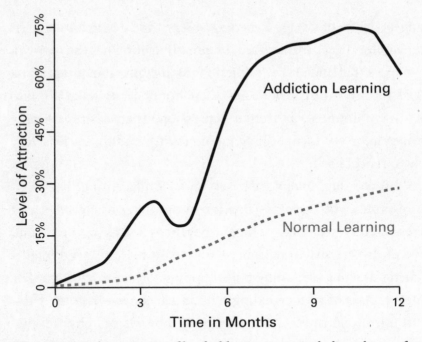

Figure 2. *Deep learning:* a profile of addiction onset, including phases of accelerated learning, stability, and reduction.

her drug-stealing storms. The learning spiral would have first quickened, showing a snowball effect in behaviour and a cascade of neural changes, when Donna and Brian began to pursue drugs for the feelings they provided, not as a means to an end—when desire kicked in and their quest for drugs overrode their other goals. That's when their lives started to unravel.

Because the onset of addiction must include one or more phases of accelerated learning, but can also simmer for long periods, I've settled on the phrase *deep learning.* This is meant to describe the overall profile of addictive learning, including periods of rapid change, periods of coasting, and temporary *remissions* (in medical parlance). Note that this profile, sketched in Figure 2, resembles a standard learning curve. It's just steeper.

Desire is at the top of the list when it comes to emotional states that propel learning. But we have to remember that negative emotions, like anxiety and shame, fuel synaptic configurations as well—configurations that strengthen themselves over development, as in the buildup of depressive or anxious dispositions. So addiction is not fundamentally different from other unfortunate directions in personality development: a self-reinforcing habit based on intense emotions, encountered repeatedly. Most important, these developmental shoots—those we see as personality changes and those we see as addiction—can sprout at the same time, each growing out of the other, supporting each other. Synaptic networks are not only self-reinforcing but also mutually reinforcing, because the brain likes to conserve structure and resources, as do all living things.

The *motivational* springboard of habit formation lies right at the centre of the brain, inside the cortex, surrounding the midbrain. This is the "motivational core" that includes the striatum, OFC, and amygdala. Changes in the wiring of these regions occur when animals are repeatedly motivated to get something or get away from something, as are hungry rodents, squawking toddlers, and horny teenagers. Much of this rewiring is the product of dopamine uptake in response to highly compelling goals, creating an ever-tightening feedback loop between desire and acquisition, wanting and getting. As to the *cognitive* aspects of habit formation, we can see at least as much rewiring in the synapses of the cortex. Biases in choosing, planning, attending to, and evaluating one's options rely on cortical changes, synchronized with activity in the motivational regions just mentioned. In short, intense emotions, focused attention, and cognitive adaptations harness each other, and together they leave enduring footprints in neural tissue. That's how brains change over the course of development, and that's how habits are formed.

So, what exactly is addiction? It's a habit that grows and self-perpetuates relatively quickly, when we repeatedly pursue the same highly attractive goal. Or, in a phrase, *motivated repetition that gives rise to deep learning.* Addictive patterns grow more quickly and become more deeply entrenched than other, less compelling habits because of the intensity of the attraction that motivates us to repeat them, especially when they leave us gasping for more and other goals have lost their appeal. The neurobiological mechanics of this process involve multiple brain regions, interlaced to form a web that holds the addiction in place. Often, emotional turmoil during childhood or adolescence initiates the search for addictive rewards, which can provide relief and comfort for a while. But there are other points of entry too. Addiction is a house with many doors. However it is approached, and however it is eventually left, addiction is a condition of recurrent desire for a single goal that gouges deep ruts in the neural underpinnings of the self.

WHY DESIRE?

It's common knowledge that addiction is characterized by strong desire and a narrowed beam of attention, which remain locked together for long periods or recur together often. This shouldn't be surprising, since strong desire always captures and holds attention. In fact, this dynamic duo has a name: attraction. A quick look at the brain shows us that desire, generated by dopamine uptake in the striatum, captures attention in the form of *expectancy*—attention to what's next, shimmering in the synapses of the OFC. The circuits connecting these structures grow like ivy as addiction takes hold. With the inclusion of the amygdala, which triggers florid emotional highlights, this neural assembly serves as the motivational engine of addiction.

But let's consider the importance of desire over other motivational states. Desire deserves a lot of respect, as does the neural terrain bequeathed to it by evolution. Think again about that forkful of pasta approaching your mouth. Notice where your attention is directed while you are eating. While the food is approaching your mouth, en route from your plate, at least some of your attention is likely to be focused on the food. For now, attention is linked with the goal of getting the food where it's going. And you are feeling desire, at least at some level. If you were not, there would be no point in eating another bite. Again, desire and attention converge into one beam. But as soon as the food is in your mouth, your attention goes elsewhere: back to the conversation or to the book you're reading or the show you're watching. The amount of attention you pay to the taste of that mouthful is a drop in the bucket compared to the amount you paid to getting it there.

So, perhaps sadly, maybe ironically, pleasure is a small part of the common experience of eating, even when you're eating something delicious. Desire and expectancy make up most of the experience: the approach is by far the main act. But this disproportionate relationship makes perfect sense. The evolutionary requirement to focus on pleasure is almost nil. Once the food is in your mouth, there's really nothing more you have to do to improve your odds of survival and procreation. But the evolutionary requirement of getting the food to your mouth, as driven by desire, is immeasurable. If you weren't deeply engaged, focused, attentive, and determined to achieve the food-in-mouth goal, your survivability would be a poor bet.

The striatum is a large, complex structure, whose purpose is to pursue and achieve goals by manipulating your behaviour with various shades of desire—impulsivity, compulsivity, need, and craving. Each of those shades may be generated by a different region, or combination of regions, in the striatum and a distinct concoction of dopamine soup. Pleasure, on the other hand, is achieved by a relatively small segment of brain (about a cubic centimetre, according to neuroscientist Kent Berridge) that appears to include a small portion of the striatum and a very different chemical recipe.

So desire is really the big wheel in all our goal-directed activities. And addiction is no exception. The critical role of desire in the brain has been the focus of research in Berridge's lab for well over a decade. Berridge was the first to argue that addiction is about wanting, not liking—desire, not pleasure—while the rest of the field has been catching up slowly. The low profile of pleasure in addiction explains why Natalie kept shooting heroin, Brian kept smoking meth, and Johnny kept drinking, long after the enjoyment dimmed to an ember of its former glory. And why smokers are rarely heard to celebrate the pleasure they get from smoking—at least after the first cigarette of the day. Even the satisfaction afforded by relief doesn't remain in attention for long. But the drive to *get* that relief, to acquire it, especially when it's been out of reach for a while, takes on colossal proportions.

Not that pleasure isn't important. There's a reason why all species of fruit have evolved to produce sugar: so that mammals will eat them and spread their seeds. Pleasure is great for triggering desire—*I want more!* But once that connection is made in Act 1, Scene 1, the audience turns its attention almost exclusively to desire.

The biology of desire not only helps us understand addiction; it helps us understand why addiction is not a disease. Why it is, rather, an unfortunate outcome of a normal neural mechanism that evolved because it was useful.

A BAD PLACE TO COME FROM

I've compared addictions to love affairs and shown that they share many psychological and neurological features. Both addictions and love affairs are ignited by attraction—highly rewarding until they start to cause more trouble than they're worth. They are both amazingly difficult to walk away from until their consequences become intolerable. And they both satisfy potent emotional needs or they simply would not be so captivating. Addictions always satisfy emotional needs. Sometimes they are generic human needs, like the need for stress reduction, comfort, pleasure, self-promotion, and the need to feel connected with something or someone outside ourselves. Since we all have those needs, any of us can become addicts. But addiction is most likely to arise from more specific needs, nestled in personalities with specific wounds, the result of hurtful or disorienting conditions in childhood or adolescence. Young people are helpless to control their conflictual or chaotic environments and the negative emotional constellations, like anxiety and depression, that result from them. In fact, the task of controlling these negative states can be impossibly difficult. Drugs and other addictive practices offer a potent antidote to what it is they are feeling, to banish their sense of helplessness and escape the pit of depression, at least for a while. That's why addiction so often follows psychological, social or physical adversity in the early years of life. The self-medication model of addiction highlights this connection, and it fits well with my emphasis on learning and development.[4]

Natalie and Donna were depressed as children. Natalie's depression was a response to a cantankerous stepfather she could not avoid or escape. Donna's depression came from muffling herself in an effort to keep the peace in a fragile, alcoholic household. Both lived with anxiety in a confusing and threatening world. Both were social isolates who relied on fake personae at home and at school. And both developed a powerful attraction to opiate drugs, whose biological and psychological contribution is a sense of comfort, safety, and reprieve. Brian was alternately berated and smothered by a mother with her own emotional problems. His childhood was marred by continual expectations of criticism and a fruitless search for reassurance. What methamphetamine offered him was a sense of strength, support, and independence, a package that neatly compensated for what he'd been missing as a child. Johnny spent his early teenage years hiding from sexual predators, riven with shame at his bed-wetting. Then he discovered that alcohol reduced his anxiety. So his fear and sense of inferiority led to a series of learning events: alcohol regulates anxiety; the more you drink, the better you feel.

Personality development is really nothing more than the laying down of habits for getting along in the world: habits of attraction, habits of self-regulation, habits for easing emotional concerns, habits for avoiding the rough edges of our lives. So we can view addiction as a branch of personality development growing out of the residue of unmet needs and failed attempts. Uncontained anxiety and depression push us to find new sources of relief. Low self-esteem sensitizes us to opportunities for feeling masterful. But once addiction becomes the main act, our inadequate resources for self-regulation have to tackle the most challenging negative emotion—craving— which is generated by the addiction itself. The addiction then becomes the unmet need that overshadows all the others, continuing to build its own callused shell. Until it's necessary to find a radically different means of self-regulation.

Evidence for the impact of early adversity comes in many forms, including a seminal study conducted in collaboration with the Centers for Disease Control (CDC) and sampling 17,000 middle-class Americans (a vast sample size). This study looked at adverse childhood experiences (ACEs) in relation to subsequent physical and mental problems. Maia Szalavitz nicely summarizes the results pertaining to addiction in a bold online journal called *The Fix*.[5] An ACE score was calculated for each participant, based on the number of types of adverse experience he or she reported during childhood or adolescence. These included physical abuse, emotional abuse, sexual abuse, alcoholism in the immediate family, and chronic parental depression (at least one of which was evident in each of the five biographies recounted in this book). The results are straightforward: the higher the ACE score, the more likely a person was to end up an alcoholic, drug user, food addict, or smoker (among other things). Two graphic examples are shown in Figure 3.

These results show that early adverse experience predicts a 500 percent increase in the incidence of adult alcoholism and a 4,600 percent increase in the incidence of IV drug use. Despite criticisms of the study, based mostly on concerns with retrospective self-reporting, these correlations are huge, they are meaningful, and follow-up prospective studies are finding similar results.

The neural picture is also straightforward. Post-traumatic stress disorder (PTSD), depression, and anxiety disorders all hinge on an overactive amygdala yoked to OFC appraisals of threat or loss and a jumpy striatum intent on relief. These components of the motivational engine get installed before young people have a chance to develop a sophisticated (and realistic) guidance system, based on maturation of the dorsolateral PFC—the bridge of the ship—which is not fully tooled until early adulthood. The traumatized amygdala keeps signalling the likelihood of threat or rejection, even when there is nothing of immediate concern in the environment. And the high whine of the whole motivational engine activates networks dedicated to finding safety, control, or relief as soon as possible. Drugs, booze, gambling, and porn take us out of ourselves. They focus our attention elsewhere. They may rev up our excitement and sense of mastery (in the case of speed, coke, and gambling), or they may quell anxiety directly by lowering amygdala activation (in the case of sedatives, opiates, booze, and maybe food). Addicts and ex-addicts know precisely how valuable these sources of succour become. We find something that relieves the gnawing sense of wrongness, we take it, we do it, and then we do it again.

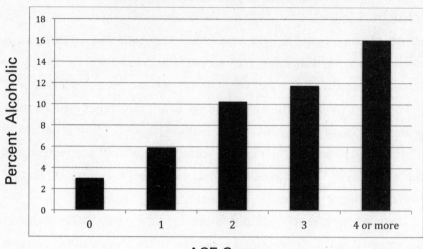

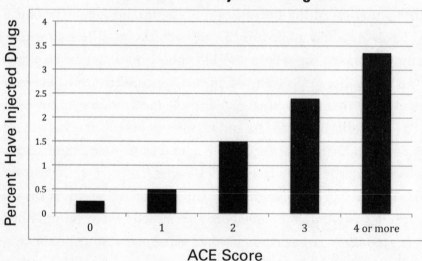

Figure 3. Graphs showing the correlation of ACE scores with alcoholism (first panel) and injected-drug use (second panel). From V. J. Felitti, "Ursprünge des Suchtverhaltens—Evidenzen aus einer Studie zu belastenden Kindheitserfahrungen," *Praxis der Kinderpsychologie und Kinderpsychiatrie* 52 (2003): 547–559.

WHAT BRAIN CHANGE MEANS FOR ADDICTS THEMSELVES

Similar brain changes take place across a wide variety of addictions, including addictions to substances and addictive activities such as gambling and bingeing. We assume that these neural events are pretty universal; they have been replicated in scientific studies around the world. But to try to glimpse what addicts feel and how they try to cope when these changes take place, we need to listen to their own stories.

The chapters recounting Natalie's addiction to heroin and Brian's addiction to meth brought home the fundamental brain change underlying all addictions: synaptic modification of the ventral striatum or accumbens, fed by surges of dopamine from the midbrain, released by cues associated with a specific reward. Dopamine uptake becomes more and more specialized with addiction, because the same synaptic network is reinforced by dopamine every time a certain substance or activity is perceived, remembered, or pursued. And further experience with that substance or activity consolidates that network, until there is a well-defined neural circuit especially tuned to that and only that reward: *Come and get it!* Meanwhile, other rewards just aren't pursued much anymore, so the networks they used to activate have gone to seed. The result is less and less competition from other neural colonies, like those tuned to parental approval, pizza, or a favourite TV series.

It's no accident that addiction and love look pretty much the same on a brain scan. The intensity of desire, the singularity of the goal, and the waning competition from other goals are common to both. And in love and addiction both, the web of connections from the striatum to other brain parts, like the OFC, allows desire to capture attention, creating and maintaining a beam of attraction. Both Natalie and Brian reported the obsessive nature of their thinking, their inability to wrench their thoughts away from drugs. They couldn't *not* think about how far away their dealer was or how soon they would meet up. Once their addictive habits were formed, coherent synaptic configurations continued to generate thoughts and expectancies about how the drug would feel and when they might have more. Images of heroin ran through Natalie's mind when she saw a bent spoon at the restaurant; Brian prepared his wake-up pipe the night before he lay down to try to sleep. Brian spoke of meth in a manner approaching worship. Natalie came home to heroin after work the way some people come home to a partner or a pet.

In the chapters about Natalie and Brian, I did not discuss compulsion as a distinct flavour of desire. (There was already enough to talk about.) And compulsion wasn't mentioned specifically as a part of Donna's struggle. But biography can illuminate the biology of addiction as much as the other way around. The stories of Natalie, Brian, and Donna suggest that compulsion generally grows out of more ordinary (but intense) forms of desire as addiction takes root over time. The pursuit of drugs evidently became compulsive for Natalie when she continued checking her cellphone every few minutes during working hours, waiting for Steve to ring. And for Brian, who felt he really had to stop at his dealer's house on his way to visit his daughter. Donna doggedly pursued her self-styled drug raids even when the risks were enormous, and the rigid constriction of her emotional world had a compulsive flavour well before she became an addict.

Perhaps compulsion should not be considered a form of desire, but rather a rigid, even mindless control mechanism. Yet, biologically, compulsion follows regularly from the goal-seeking activity of the accumbens. With repeated goal pursuit, the maypole of dopamine fibres from the midbrain winds its way along the striatum until it arrives at its dorsal (northern) summit. And then behaviour is no longer driven solely by the expectancy of reward but also, and often primarily, by the intent to satisfy an urge. Johnny's drinking habits epitomized the compulsive nature of late-stage addiction. There was little pleasure and a great deal of suffering—a kind of self-imposed torture—awaiting him every time he got beyond his first two drinks. And Alice's dedication to self-starvation, followed by binges that deeply shamed her and threatened her marriage, were compulsive in the extreme. These are the unsavoury aspects of brain change that nevertheless follow the classical rules of human learning.

Once the striatum has established its dominion over the dopamine pump, the motivational engine is turned on and let loose. Then habits of desire settle into place and the rest of the brain adjusts. In the chapter about Donna, the medial prefrontal cortex (medial PFC) was the brain region in the spotlight. Like many other neural portrayals in this book, that was a bit of an oversimplification, but it made a point. Brain change in addiction concerns more than our perception of what we desire; it also underlies changes in how we perceive ourselves and others, how we define and fashion our identity. As Donna negotiated her addiction she also perfected her ability to lie, to take satisfaction from deception, and even to rationalize and accept who she had become. The changes in her medial PFC seemed to support two distinct personality patterns: one focused on pleasing others, the other a hotbed of resentment and defiance well served by stealing and lying. We can envision these pathways remaining separate during her addiction and then converging during her recovery. The point is that addiction is a phase of individual development, not only in the addictive habits themselves but also in the transformations shaping the person as a whole. There is no way to comprehend addiction outside of development, and no way to comprehend people's development as distinct from their addictions.

Two unique psychological mechanisms make addiction particularly tenacious. In the chapter about Brian, we saw how now appeal (delay discounting) rivets attention to immediate rewards and devalues future gains. The chapter about Alice highlighted ego fatigue, the breakdown of cognitive control when people try to suppress their feelings or block their impulses for some length of time. Both of these psychological vulnerabilities are natural, both are shared with other animals, and both correspond with specific neural events. Now appeal was famously depicted by Walter Mischel's marshmallow test: three-to-four-year-old children were told they could eat one marshmallow as soon as the examiner left the room, but if they waited a few minutes until she returned, they would have two to eat. Three-year-olds most often gobbled the first marshmallow in no time, thereby losing the opportunity for a second. Four-year-olds were better at waiting. But a look at the video records of this experiment (or its many derivatives) reveals ego fatigue as well. Many children, and many adults, can resist the temptation for only so long. After a minute or two of staring at, even fondling and licking, the marshmallow in front of their noses, they lose their self-control and surrender to desire.

Now appeal consists of a narrowed beam of attention toward a highly attractive and imminently available reward. That is precisely the state addicts find themselves in day after day, time after time. And the culprit, again, is striatal dopamine. One of dopamine's chief concerns is to highlight available goals. Immediate goals are available goals, and striatal networks surge with dopamine whenever those goals announce, *Here I am!* The wash of dopamine disperses the appeal of alternative goals, like leaving enough money to pay the babysitter or being a better person a week from Tuesday. But addicts who have not surrendered completely to their addiction—and most have not—enlist the help of their prefrontal cortex to resist the temptation of the immediate. And they can resist, for a while. They resist by applying the techniques they've been taught, by counsellors, therapists, judges, parents, and just about everyone else. Resist! Inhibit! Yet the research tells us unambiguously that suppression is the wrong way to go, because it accelerates ego fatigue. The best way to resist temptation is to shift perspective and reinterpret your emotional state. Instead of tying yourself to the mast in order to resist the Sirens' song, you must recognize the Sirens as harbingers of death and reframe their songs as background noise.

Of course, children eventually learn the cognitive controls necessary to resist temptation. In fact, they learn to look beyond now appeal and stiffen their resistance to ego fatigue more or less in parallel. We could say that their willpower simply improves with age, but that doesn't tell us much. More informative are the brain changes that support improved self-regulation as children grow up. The prefrontal cortex changes massively during childhood and adolescence. One way to describe this trend is to say that the more ventral regions (like the OFC) mature first, and the more dorsal regions (like the dorsolateral PFC) become wired up over a much longer time span. As noted previously, the dorsolateral PFC is among the latest regions to mature, and its capacity for insight and judgement remains a work in progress until the early twenties. That may be why the more sophisticated forms of self-direction, like reevaluating one's goals and placing them in perspective, must wait until the adolescent years are behind us. The higher echelons of self-control count as regal achievements in the synaptic wiring of the brain. And, like other achievements in neural development, they rely in large part on practice—which takes time.

LOSING CONTROL

Research shows that children's ability to overcome now appeal improves with age from middle childhood to middle adolescence. And at least one study attributes that improvement to the maturation of the (left) dorsolateral PFC.[6] Quite a number of other studies have shown that adults control now appeal by activating the dorsolateral PFC, often on the left side of the brain. One of my favourites was conducted by Bernd Figner, now at Radboud University in Nijmegen, the Netherlands, where I currently work.

Figner used transcranial magnetic stimulation (TMS), a procedure that temporarily disrupts activity in the cortex, to look at cortical mastery over people's vulnerability to now appeal. As shown in Figure 4, he placed the TMS machine over the left or right dorsolateral PFC of his adult participants and had them sit in front of a screen. Slides on the screen offered a series of choices between immediate rewards of lower financial value and future payoffs of higher value. Participants used a button-press pad to select their preferred choices. When the device was placed on the left side of the head and the current was turned on, participants chose immediate rewards more frequently than they had a few minutes earlier. There was no such effect when the device was placed on the right side of the head. In other words, they needed their left dorsolateral PFC to overcome now appeal and choose those future gains. When that region was partly shut down, their interest turned to whatever was more immediate.[7] I'll get back to the left-right issue in a few more paragraphs, but first it's important to understand the role of the dorsolateral PFC in the top-down control of impulsive behaviour more generally.

The biographies of Johnny and Alice tell the story of a brain that continues to change with addiction, not only in the motivational core but also in the dorsolateral PFC—the bridge of the ship. How should we think about these changes? The dorsolateral PFC becomes overactivated in the early stages of addiction: *This is really rewarding. I like it a lot, and I want to keep doing it. But I should control it, shouldn't I? (Otherwise I'll probably get addicted! Or go broke . . .) So I'm only going to snort it, not shoot it. Or I'm only going to drink at night, not during the day.* But after a while,

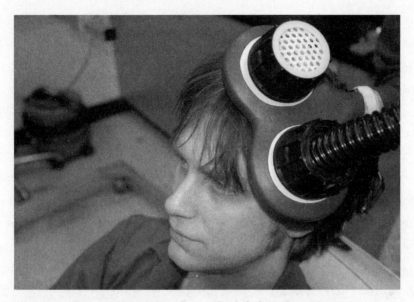

Figure 4. TMS machine placed over the left dorsolateral PFC. Courtesy of Timothy Spellman, and of Bernd Figner, Behavioural Science Institute and Donders Centre for Cognitive Neuroimaging, Radboud University, Nijmegen, Netherlands.

with a variety of substances and some eating disorders (including binge eating), the dorsolateral PFC becomes partially disconnected from the striatum. The bridge of the ship loses its command over the motivational engine. And now consumption becomes a reckless child who will not listen. The reasons for this disconnection are complex and not fully understood. But suffice it to say that habits free themselves from higher-order controls because the striatum no longer sends out requests for prefrontal engagement. This would be around the time that compulsion takes over from impulse, so we can think of control as switching to the dorsal striatum, the supervisor with the lowest possible IQ.

Once addiction has set in like a thick fog, top-down control goes missing whenever your dealer calls or the liquor store is about to close. The motivational engine continues to roar while the captain on the bridge goes back to his poker game. Functional connections are lost, which means some of the synaptic pathways get pruned and eventually disappear. Now structural connections are lost. This explains the loss of grey matter volume reported with long-term addiction.

As noted previously, synaptic pruning is a normal developmental process. It does not indicate disease. In fact, research shows that, when the same inputs are encountered repeatedly, some connections are depleted in order to improve overall efficiency.[8] But addicts are hard pressed to overcome now appeal once the bridge of the ship stops responding. Like the participants in Figner's study, they find that their dorsolateral PFC has taken a coffee break, leaving them more susceptible to urges best ignored. Little do they know that their striatum has stopped sending the right signals. This state of disengagement, while not abnormal in the brain's game of shape-shifting, is highly unfortunate for people trying to resist the appeal of immediate goodies, be they Internet games or crack cocaine. If the dorsolateral PFC is necessary to overcome now appeal, and that brain region takes a break whenever addictive opportunities present themselves, then self-restraint must now rely on more primitive methods of control. Raw inhibition and suppression (bequeathed by more ventral regions of the PFC) are all they can come up with. In other words, to resist the dictatorial commands of compulsion, broadcast from the dorsal striatum, addicts are forced to rely on control strategies that are essentially childish. *I won't do it! I won't! I won't even think about it!* (Or in the famous words of the War on Drugs: *Just say no.*)

In fact, this is the worst possible recipe for self-control, because it actually hastens ego fatigue. Alice described her attempts at staving off her binges as trying not to think of an elephant. Not very effective. Now recall that ego fatigue *also* corresponds with a disconnect between the motivational core and the dorsolateral PFC, though that disconnect is functional, not structural: it transpires over minutes, not months. So ego fatigue weakens contact between impulse and self-control just as addiction does, and both are underpinned by the same neural changes, though they take place over vastly different time scales psychologically and biologically.

To summarize, now appeal strengthens impulses and crosses the line into compulsion, perpetuating habits of desire based on the immediate. Meanwhile, ineffective methods of self-regulation accelerate ego fatigue, releasing these habits from what remains of dorsolateral supervision—the bridge of the ship. Over time, the wiring diagram of the prefrontal cortex shifts toward constellations more typical of kids than adults. Ego fatigue wins the point, and now appeal wins the game. These brain changes are formidable, and they can be reinforced by the loss of alternative synaptic pathways, through ongoing pruning. But they are rarely permanent. Synaptic pruning is a means to an end, as the brain continually updates its repertoire. The loss of grey matter volume in a few cortical regions makes the brain more efficient, more streamlined—though we might not approve of the particulars—much like the massive cortical pruning that goes on in early adolescence. Yet addiction can be a brick wall or a cliff edge; changes in the dance between desire and self-control spell suffering for many and death for some. In the next chapter I'll explore how the sometimes tragic consequences of long-term addiction can be avoided, by redirecting rather than extinguishing the biology of desire.

WHAT'S LEFT?

To conclude this chapter, I'm going to shift from the established (and replicated) findings of addiction neuroscience to more speculative terrain, and suggest a refinement to the neural map of addiction I've drawn so far. Most of us have been taught that the two hemispheres of the brain—left and right—are distinct in their functions. Each has its own perspective, its own code, its own jobs to do. There is no doubt that this is true, yet I haven't mentioned hemispheric specialization yet. Things were complicated enough. Now let's split the cortex down the middle and think about what each side contributes, or fails to contribute, when people become addicted.

Focusing on the prefrontal cortex, we can identify the left hemisphere with analytical thinking, planning, thinking in sequences, and language. In fact, language might be the necessary code for all these "linear" forms of thought. A stroke or injury in the left PFC makes it difficult or impossible to speak or to gesture in any meaningful way. It leaves us in a private world without recourse to normal communication. It also makes it difficult or impossible to plan, judge, or act in a rational and organized fashion. We need our left brain to help us stay on track and optimize outcomes. In contrast, the right PFC does not specialize in language. It gets "meaning" in a more spontaneous way, based on information processed globally, often spatially, rather than piece by piece from the words in a sentence. The right hemisphere is often considered not only more intuitive but also more "emotional"—perhaps because it cannot divert the awareness of feelings the way the left can by thinking through situations or rationalizing challenges. Since the right hemisphere does not think sequentially, its sense of time is sometimes thought to be cyclical rather than linear. It may specialize in repeated patterns, like the passage of one day to the next or the rotation of the seasons, rather than projecting a time course that moves from past to present to future.

We have seen that addiction leads to a disconnect between the motivational core (striatum, OFC, amygdala), where desire is orchestrated, and the dorsolateral PFC, the bridge of the ship, responsible for discrimination, judgement, and conscious self-control. We have also seen that the pair of Achilles' heels most exposed in addicts consists of now appeal (delay discounting) and ego fatigue, and both these saboteurs arise from a similar pattern of disconnection—a relative loss of engagement of the dorsolateral PFC. To complete the picture, the studies just reviewed specify the *left* dorsolateral PFC as the main defence against now appeal, slow to develop in childhood and adolescence but steadfast and committed when adults forgo immediate rewards. It would not be surprising if it's the left dorsolateral PFC that begins to fizzle in extended addiction. This delineation has shown up in a number of studies, but it hasn't yet been tested rigorously. If it proves to be valid, how might it help us explain addiction?

It might help us explain the challenge faced by addicts in overcoming now appeal, based on their inability to maintain a linear sense of time. Future well-being can only be pursued if the future can be imagined. And to imagine the future requires a linear sense of time—time that moves forward like an arrow. But the linear sense of time maintained by the left hemisphere may become porous or amorphous once the dorsolateral PFC on that side of the head is numbed by a quest for the immediate. Once addictive goals are by and large the only goals being sought, there may be little to look forward to, and little *ability* to look forward to anything, beyond what's going on in the present. Yet we need something special, some connection between now and later, to get us past the enchantment of the immediate to the possibility of a better future. That something is sometimes called *will*. In fact, George Ainslie's masterful 2001 book about delay discounting and its role in addiction is simply entitled *Breakdown of Will*. That breakdown is the essential failing that delivers us to the cheap thrills of the immediate. And it may well reflect a corresponding breakdown in communication between the left PFC and the motivational core.

Philosophers and neuroscientists argue relentlessly about the existence of free will, but we all know what it feels like to hold out for something better than what's presently available. We know that it takes not only effort but also vision or insight—the capacity to stretch our sense of ourselves to the possibility of something that doesn't yet exist. The dorsolateral PFC is the only part of the brain sophisticated enough to imagine that future, evaluate its worth, and decide to pursue it. The gift of the left brain may be a dorsolateral PFC with the power to think in straight lines—the power to think ahead.

The motivational core of the brain needs to connect to many other regions in order to live up to its evolutionary potential. But of all the regions that can help us negotiate the maze of attractions in this complex world, the higher reaches of the prefrontal cortex, with their exquisite capacities for comparison, evaluation, judgement and choice, are the most essential. Connections between the limbic structures of motivation and the dorsolateral PFC generally develop and stabilize, not only through adolescence and early adulthood, but also throughout life, whenever new treasures dazzle us. Except that addiction modifies those connections, casts them in shadow, dampens them like kindling that is no longer quick to light. As long as addiction is tolerated, that reconfiguration of synaptic patterning remains. And then, when it can no longer be tolerated, the brain changes again, reconnecting its motivational engines with the bridge of the ship, from where we can see the horizon.

NINE

Developing Beyond Addiction

T here's a good deal of debate about how to promote recovery from addiction. Medically oriented professionals see pharmaceutical agents as the best possible defence. Presently available drugs can suppress cravings, ease withdrawal symptoms, and combat concurrent problems such as depression. However, pharmaceuticals are far from foolproof. They work only as long as they continue to be taken regularly, and some of them actually prolong the addiction, though in a slightly altered guise. Other approaches to recovery can be summarized, very broadly, as based on one of two lines of thought: either "just say no"—that is, grit your teeth, suppress urges, and maintain your traction—or cultivate openness, insight, and perspective change. Real-life recovery probably involves both tactics at different stages, but if I had to put my money on one as essential, I'd pick openness leading to perspective change.

I believe that getting past one's addiction is a developmental process—in fact, a continuation of the developmental process that brought about the addiction in the first place. And the biology of neural change—the way brains transform themselves and the way habits form and reform—helps explain how that developmental process works. In fact, the importance of ongoing brain change becomes difficult to dispute when we link the neurobiology of addiction with the stories of those who have been there and moved on. That's the insight that inspired the structure of this book.

NEUROPLASTICITY

The term *neuroplasticity* has been bandied around a lot. Norman Doidge and other authors have helped bring it into the public spotlight, but it's been recognized as a fundamental characteristic of the human brain for well over a hundred years. Dr. Eric Kandel of Columbia shared a Nobel Prize in 2000, reflecting decades of research on how the brain changes when learning occurs. In a nutshell, Kandel showed that the connections among neurons—the synapses—must change physically if memories are to be formed. He showed this at a molecular level, validating Hebb's famous insight from the 1940s: "what fires together wires together." Both learning and memory rely on that basic formula. The neural changes that take place when we learn and remember things embody neuroplasticity. Neuroplasticity is the brain's natural starting point for any learning process. This includes the development of addiction, of course. But it's also the springboard to recovery.

Neuroplasticity is strongly amplified when people are highly motivated. Which is why all learning requires some emotional charge, and why entrenched habits like addiction grow from intense desire. Clearly, the desire to recapture a potent experience of pleasure or relief is the motivational on-ramp to addiction. But does motivation also foster recovery? Is the high beam of desire necessary to find the road out?

In *The Woman Who Changed Her Brain,* Barbara Arrowsmith Young describes the many cognitive exercises she devised for herself in order to overcome her very severe learning disabilities. She practiced these exercises prodigiously. As a result, she went from a high school student who could not comprehend history or decipher simple sentences to a writer and teacher who has set up roughly seventy schools for learning-disabled children across North America. I met this remarkable woman in Australia, at a book fair, and I became convinced that her intuition, creativity, and determination to triumph over her learning disabilities were precisely the means by which addicts recover. I also learned her delightful phrase for the neuroplasticity needed to replace bad habits with good ones: *what fires together wires together, and what fires apart wires apart.* In other words, new mental patterns often require diverging synaptic traffic routes—two roads to Rome where there had been only one.

In 1993, Alex Mogilner and colleagues looked at the brains of people plagued with "webbed" fingers (Figure 5). That means that some of their fingers were connected and could not be moved independently—they could only operate in unison. After surgery was performed to allow the fingers to move separately, these authors looked at changes in the somatosensory cortex. What they found was that clusters of neurons that had always fired together now fired partially independently: "The presurgical maps displayed shrunken and nonsomatotopic hand representations. Within weeks following surgery, cortical reorganization occurring over distances of 3–9 mm was evident, correlating with the new functional status of their separated digits."[1] So the brain adjusted its wiring, breaking down the coherent habit it had assumed, based on the (inescapable) details of repeated action patterns, and replacing it with new habits, based on novel action patterns. In fact these changes were observable just weeks after the change in action patterns took place. Might recovery work the same way?

Just in case you think the webbed-fingers analogy is farfetched, you should know, first of all, that it's not an analogy. That's exactly

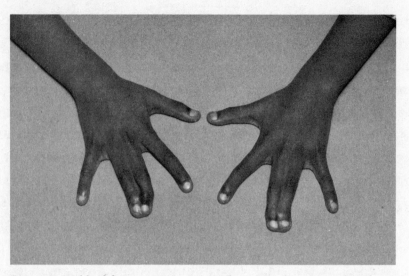

Figure 5. Webbed fingers. From A. Mogilner, J. A. Grossman, U. Ribary, M. Joliot, J. Volkmann, D. Rapoport, R. W. Beasley, and R. R. Llinás, "Somatosensory Cortical Plasticity in Adult Humans Revealed by Magnetoencephalography," *Proceedings of the National Academy of Sciences* 90, no. 8 (1993): 3593–3597. Courtesy of St. Bartholomew Hospital/ Science Photo Library.

how neuroplasticity works, whether it's dealing with severe learning problems (which no doubt involve the prefrontal cortex) or reversing a physical anomaly that emerged during prenatal development. Second, the plasticity in the webbed-fingers example is more likely to underestimate than overestimate the plasticity of the dorsolateral PFC, the region responsible for conscious decision making, judgement, and deliberate self-control. The wiring of the motor cortex is completed by middle childhood, not surprisingly, since most action sequences have been well learned by that age. If the motor cortex can reorganize (at least in its details) within a few weeks, long after it has matured in childhood, imagine the potential for reorganization in the dorsolateral PFC, a region that remains plastic for decades.

When people recover from strokes or concussions, the same sort of rewiring takes place in many regions of the cortex. Even language, one of the most basic human functions, can be relearned after it has been demolished by brain damage, through the synaptic rewiring of cortical regions that previously took care of other business. For example, following a stroke to language areas in the left hemisphere, the right hemisphere will often increase its participation in language functions, greatly facilitating language recovery. Thus, neuroplastic change can and does occur in real life, with a speed and vigour we rarely imagine.

Back to addiction. People *learn* addiction through neuroplasticity, which is how they learn everything. They maintain their addiction because they lose some of that plasticity. As if their fingers had become attached together, they can no longer separate their desire for well-being from their desire for drugs, booze, or whatever they rely on. Then, when they recover, whether in AA, NA, SMART Recovery, or standing naked on a thirty-third-floor balcony of the Chicago Sheraton in February, their neuroplasticity returns. Their brains start changing again—perhaps radically. As we can see in each of the five biographies, *they start to separate one set of desires from another and to act on them independently.* For example, Johnny diverted his desire for alcohol into a quest for bodily relaxation that could be satisfied with yoga. Alice separated her preoccupation with food into habits she labelled as healthy and unhealthy. Just as in Mogilner's study, their brains began to grow new synaptic patterns to allow for those distinctions, hold on to them over time, and thereby acquire new vistas of personal freedom and extended well-being.

The take-home message here is simple: recovery involves major changes in thought and behaviour, and such changes require ongoing neural development. Without developmental adjustments in synaptic patterns, we would stay exactly the way we are. Which raises a question: why should we call it "recovery" if it's the beginning of something new? (I want to emphasize that the term "recovery" has become emblematic for addicts overcoming their addictions, and many individuals and organizations use the phrase "in recovery" to denote people's success in moving on. Although I think the term is ill-founded because of its medical connotations, I have great respect for its use in reference to quitting addiction, whether by individuals or by treatment programs, rehabs, or support groups that assist in this process.)

EXPERIENTIAL AND NEURAL PATHS TO QUITTING

The way people become addicts, and the way they move beyond their addictions, is complicated and mysterious; it can't be captured by a single word like "neuroplasticity." During periods of addiction, brains are remarkably stuck. The most candid thing one can say about neuroplasticity during such times is that it's nowhere to be found. What makes the difference? How do people move from stuckness to flexibility once their self-destructive neuronal habits have been rehearsed thousands of times? In this section I note some of the psychological stepping-stones that seem necessary to overcome addiction—suggested by the experiences of those who succeeded—and blend in their probable neural counterparts. Here our understanding of brain change must not only inform but also follow from insights gathered from individual experiences.

But before recounting those experiences, I want to summarize the neuropsychological mechanisms I've highlighted as the cornerstones of addiction: now appeal in relation to narrowly defined rewards; ego fatigue arising from attempts to resist temptation; and personality development, the ongoing stream of change and stability in individual minds.

The role of now appeal in addiction is critical. We mammals often find immediate rewards more attractive than future rewards of higher value. But now appeal is driven by dopamine uptake to the striatum:

immediate rewards are always more compelling than long-term rewards, and dopamine circuitry evolved to capitalize on that simple fact. In addiction, this mechanism narrows in on itself insidiously. When addictive rewards colonize dopamine circuitry, and long-term rewards become increasingly hard to recall, addictive rewards become the only game in town. Then what? Then you know you're in trouble, so you try to resist temptation and inhibit the recurring desire for *more*. Which leads to ego fatigue. Ego fatigue is like a microcosm of addiction: a loss of top-down cognitive control, augmented rather than diminished by attempts to suppress impulses. Ego fatigue reflects a disconnect between the bridge of the ship—the dorsolateral PFC—and its motivational engines in and around the striatum. Most important, the same disconnect, the loss of communication between these brain systems, becomes entrenched through synaptic pruning when addiction crystallizes over months and years.

Failures in recovery—and so-called relapses—can easily be explained by the exhaustion of self-control, when now appeal and ego fatigue work together and the bridge of the ship becomes harder to access from the motivational engine room. Attempts to suppress the attraction of immediate rewards amplify ego fatigue, so we give in to desires we might otherwise circumvent. Yet addiction takes time, carves its ruts deeply, and builds on existing mental habits. So addiction is an aspect of personality development, often an extension of patterns formed by failed attempts to deal with negative emotions in childhood. In fact, addictive patterns aren't just an outcome of personality development; they're a continuation of that development. Addicts continue to grow and modify their self-image, generally for the worse, and so addiction becomes a part of the self, characterized by the imprint of earlier failures.

These processes work together to fashion addictions. But how do they shift—how does their relationship change—when people move beyond their addictions?

When Natalie was stuck in jail and beset by boredom and despair, she had little else to do but examine how she'd arrived there and imagine possibilities for a happier future. Her desires had been devoted almost entirely to finding heroin, often daily, to relieve her anxiety and depression. This trance-like state, focusing only on the now, thwarted her ability to see her life in perspective. She couldn't see that heroin served the same purpose as escaping to her room and burying herself in books had in childhood. The past had never been thoroughly examined; the future wasn't even on the table.

And then she learned to meditate. She began to see her present self in the perspective of her past: all those lonely years, shut away in her room, trying to avoid her stepfather's temper and the social exclusion she expected from others. And she saw, for the first time, that heroin had been a solution to a long-standing problem—as flawed a solution as shutting her door in childhood. That insight came with a shift in perspective that diminished the appeal of the drug itself. And it must have relied on a reconnection between the dorsolateral PFC, responsible for insight and judgement, and the motivational circuits shaped by drug seeking. She didn't put all the puzzle pieces together at once, but she started to arrange them while still in prison. Enough to see heroin for what it was—to her. Enough to avoid it, to be repelled by it, once she returned to the everyday world.

Brian began to see his own development in perspective once he started psychotherapy. He began to recognize his infatuation with meth as the outgrowth of a child's need for independence and mastery. Methamphetamine conferred a sense of competence he had never acquired from the people in his life. He learned to see his present as an extension of his past, and he stretched that insight to include a future that captivated him and propelled him to the next stage of his development.

During the years of his addiction, Brian was a poster child for now appeal. The slightest hint that he was starting to come down would send him lunging for his pipe. Like Natalie, this trance-like present tense blocked the conduit between his past and his future, making it impossible to perceive the larger story of his life from the chaos of his momentary existence. Then, with the help of his therapist, he began to construct a bridge from now to then, piece by piece by piece, leading to a future that was safe, satisfying, and valuable to himself and to others. A goal he had dreamed of in the past, now attractive enough to pursue in the present. That kind of reflective insight could only be achieved by regrowing pathways between his dorsolateral PFC and his striatum. He had to think, imagine, compare, and decide what to do *while* feeling the familiar heat of desire. Only when both systems were activated simultaneously could new connections sprout.

That's how the brain works. What fires together wires together. And what fires apart wires apart. New and original avenues are created when unexplored plans for long-term satisfaction are brought to mind, held in mind, and exposed to the forge of desire—a detour from the familiar road leading to immediate gratification. So began Brian's work toward becoming a healer rather than a victim— taking care of animals, then people—starting a therapeutic community for addicts in the countryside, and eventually opening up his own treatment program in Cape Town.

It appears that Natalie and Brian began to outgrow their addictions when they were able to reflect on their lives, connect their past to their present conundrum, and imagine a future very different from the present. I believe that this process of reflection and perspective taking was precisely what helped them overcome the now appeal of drugs. As a result, they no longer had to fight their impulses with the same exhausting effort hour by hour and day by day. This greatly reduced ego fatigue, which was the key to changing momentary behaviour and learning to rely on top-down control. They could now begin to mark their progress toward acquiring a sense of safety, born of self-care, and a sense of purpose, resulting from accomplishments that could only take shape over time. They could detect a shift in the course of their own personality development: the emergence of new mental habits, new habits of behaviour, and a different sense of who they were as people.

Until addicts begin this kind of reflection, their preoccupation with the immediate is locked in by dopamine's tide. They are caught in a rapidly cycling spiral of desire and acquisition, wanting and getting, getting and losing. Now appeal is not just a devaluation of future goodies; in the case of addiction, it's a gorge carved across the continuum of a person's life, dividing personality development into an unexamined past and an unimagined future. It's a place where people can live for a long time. In fact, they can die there. Or they can find a way out. That requires connecting the jagged halves of one's life—past and future—by discovering and recounting the story of one's own development, and extending that developmental route beyond the next few hours, into the months and years ahead.

A similar transformation seemed to occur for the other people whose biographies I've sketched. Donna discovered the enormous investment she'd made in duplicity. She learned to recognize the no-man's-land she'd inhabited between power and submission, between fulfillment and desperation. This was a key component of the therapy she pursued—the impetus for her to share unexamined secrets with someone she hardly knew. Johnny took advantage of residential treatment, AA, and finally psychotherapy to quit drinking for good. And while following these stepping-stones, he unearthed the anxieties he'd been escaping until then. Detox was only the first phase of his recovery. What kept him sober was the insight he gained about who he was, who he'd been as a child, and who he could still hope to be without the balm of alcohol. Alice's story was more complicated. She is someone who could not live without an established formula for self-control. Yet her rigid, self-constricting controls had only compromised her, leading her into the dangerous territory of anorexia and then rebounding in great destructive leaps between dieting and bingeing. In a support group for women with eating disorders, Alice found a cord that extended from her impossible present to a possible, engaging, and satisfying future. Once that cord was envisioned, pulled taut in her perspective, it became feasible to direct her life along its length.

CREATING A NARRATIVE FROM PAST TO PRESENT TO FUTURE

I want to share with you the results of an original and deeply moving research program conducted by Michael Chandler and his associates at the University of British Columbia.[2] The researchers canvassed Native communities through much of western Canada. What struck them almost immediately was the astounding suicide rate among teenagers—500 to 800 times the national average—infecting many of these communities. But not all of them. Some Native communities reported suicide rates of zero. "When these communities were collapsed into larger groupings according to their membership in one of the 29 tribal councils within the province, rates varied . . . from a low of zero (true for 6 tribal councils) to a high of 633 suicides per 100,000." What could possibly make the difference between places where teens had nothing to live for and those where teens had nothing to die for?

The researchers began talking to the kids. They collected stories. They asked teens to talk about their lives, about their goals, and about their futures. What they found was that young people from the high-suicide communities didn't have stories to tell. They were incapable of talking about their lives in any coherent, organized way. They had no clear sense of their past, their childhood, and the generations preceding them. And their attempts to outline possible futures were empty of form and meaning. Unlike the other children, they could not see their lives as narratives, as stories. Their attempts to answer questions about their life stories were punctuated by long pauses and unfinished sentences. They had nothing but the present, nothing to look forward to, so many of them took their own lives.

Chandler's team soon discovered profound social reasons for the differences among these communities. Where the youths had stories to tell, continuity was already built into their sense of self by the structure of their society. Tribal councils remained active and effective organs of government. Elders were respected, and they took on the responsibility of teaching children who they were and where they had come from. The language and customs of the tribe had been preserved conscientiously over the decades. And so the youths saw themselves as part of a larger narrative, in which the stories of their lives fit and made sense. In contrast, the high-suicide communities had lost their traditions and rituals. The kids ate at McDonald's and watched a lot of TV. Their lives were islands clustered in the middle of nowhere. Their lives just didn't make sense. There was only the present, only the featureless terrain of today.

I've thought about this research countless times since I began to study addiction and recovery. To me the message is clear: humans need to be able to see their own lives progressing, moving, from a meaningful past to a viable future. They need to see themselves as going somewhere, as characters in a narrative, as making sense. In addiction, the relentless preoccupation with immediate rewards carves a small burrow out of the potential richness of time. In psychological terms, this is the outcome of recurrent cycles of now appeal and ego fatigue. But the details are more damning. The fine print tells the story of a tightening spiral of desire, acquisition,

and loss, the narrowing of perspective and meaning, the rigid infatuation with reliable, available, but terribly boring rewards—
rewards that leave only emptiness and craving in their wake. The
addict's life is lived in the tomb of the present, dead because it has
lost its connection with the story from which it came.

In neural terms, as we have seen, this state corresponds to the
breakdown of communication between the motivational core of
the brain (the striatum, OFC, and amygdala) and the bridge of the
ship (the dorsolateral PFC). By virtue of this disconnection, desire
drives behaviour in small redundant circles, independent of insight,
perspective, and higher-order goals. Desire roars with immediacy,
craving fulfillment, but its natural partners are judgement and direction, planning and perspective, capacities programmed into the
dorsolateral PFC throughout childhood and adolescence. Once
that partnership has come apart, it needs to be put back together.

Chandler's research has another implication for addiction, and
I note it here as a suggestion for further thought. Several scholars
or experts on addiction have proposed that confining social conditions and alienation from one's culture are major ingredients in
the rise of addiction. Bruce Alexander, who conducted the Rat Park
studies (reviewed in Chapter One), has outlined a broad social theory of addiction in his 2008 book, *The Globalization of Addiction,*
where he investigates societal dislocation and its negative impacts.[3]
Carl Hart (also introduced in Chapter One) showed how members
of minority groups living in the inner city *chose* to take drugs because other choices seemed unavailable or meaningless. Clearly
these populations have something in common with Chandler's
high-suicide communities—the loss of a viable cultural narrative in
which their individual narratives made sense.

Yet a sense of dislocation is not necessarily terminal. In fact,
the personal isolation characteristic of addiction has been a prime
target for support groups fighting addiction. From therapeutic

communities to meetings in church basements, the support and
encouragement provided by other recovering addicts help people
contextualize their own addiction, through the sharing of life stories and the mutual striving for a better future. In this way, communication and consensus help build a sort of culture around the
struggles faced by addicts. It seems that the creation of this larger
narrative is intended to build a sense of continuity that works as
scaffolding for individual narratives, and that's what made the difference in suicide rates among Native youth.

FINDING A FUTURE: SELF-NARRATIVE AND SELF-TRUST

The facility for viewing one's life as a narrative may be what's missing in addiction. And the loss of an accessible self-narrative corresponds with clues that the dorsolateral PFC becomes partially
disconnected from the motivational core, both in episodes of now
appeal and over the long-term course of addiction. My focus on the
left dorsolateral PFC, although partly speculative, can help make
sense of what goes wrong when people seem unable to quit. Not
only are memories and ambitions difficult to access, but the sense
of time as a linear dimension, connecting now to later, is replaced
by a sense of time as cyclical—the right hemisphere's proclivity. Instead of a future stretching out ahead, addicts can only imagine the
reiteration of the present. If this is an accurate picture, then reconnecting the left dorsolateral PFC with the motivational core would
allow desire and perspective to work together, and that might be
the best way, in fact the only way, to build a road from the present
to the future.

Addicts experience something breathtaking when they can
stretch their vision of themselves from the immediate present back
to the past that shaped them and forward to a future that's attain-
able and satisfying. It feels like shifting from momentary blobs of

experience to the coherence of being a whole person. It feels like being the author and advocate of one's own life. It feels like being real.

Now imagine what that means for the capacity to trust one's own judgements, values, instincts, and attainments. From making choices that are obviously self-destructive, there is a shift to making choices that are self-enhancing and self-sustaining. The value of this transformation cannot be overstated. Addicts can live for years without experiencing a kernel of self-trust. Why trust that you will actually be different when the evidence suggests that you'll go on being the same? Why believe that you can pursue what's beneficial rather than what's immediately available, when you've bypassed that junction a thousand times?

To experience a sense of continuity between me now, me then, and me in the future is precious. But when it's been missing for a while, perhaps for one's whole life, it's not easy to find. It requires a perspective that can only be obtained by addressing the future in the context of the past. And it requires one other thing, one fundamental resource: desire itself. There's no way to reach forward with determination and hope unless you want badly to get there.

REALIGNING DESIRE

The importance of insight and perspective taking has not been ignored by the treatment world. Cognitive behavioural therapy, mindfulness-based relapse prevention, motivational interviewing, and the treatment programs that utilize some combination of these techniques all aim for some brand of self-reflection and perspective taking. The twelve steps of AA also encourage a major shift in perspective and the determination to maintain that shift. Approaches based on insight therapy are explicitly concerned with extracting meaning from the past to help people understand their present struggles. Yet none of these approaches fully grasps the way time collapses in addiction, nor the critical importance of reframing the links between past, present, and future.

In my view, the nature of perspective taking can't be fully appreciated or understood without considering the mechanics of now appeal, whereby addictive goals pump dopamine directly into the motivational core of the brain and time contracts to the immediate present. Thus addiction neuroscience points to an intimate relationship between perspective, now appeal, and desire. Neuroscience shows that now appeal gets its power from the biology of desire. There may be no wedge that can pry desire away from addictive goals once the dopamine pump is under their control, especially when other goals have lost their apparent value or availability. So, if desire cannot be turned off or seduced away from addictive goals, immediate goals, then it has to be fastened to goals incompatible with addiction—goals such as freedom from suffering, achievement of life projects, access to loving relationships, and the sense of coherence and self-love that can come with abstinence. And if those goals can't be envisioned, because of a static preoccupation with the present, then self-narrative and desire need to be packaged together—self-narrative to shift perspective to long-range goals, desire to power the pursuit of those goals.

Stories don't work without emotional themes. They would be impossible to follow. There would be no impetus to get to the next paragraph or the next chapter, either for the main character or for the reader. In the same way, real-life narratives can't be held in mind without the feelings that give them shape. Desire is the most potent of those feelings. Desire has the power to propel us through life, to get us from now to later. The trick to overcoming addiction is thus *the realignment of desire,* so that it switches from the goal of immediate relief to the goal of long-term fulfillment. Like the brain changes that follow the separation of webbed fingers, synaptic routes from the motivational core must activate two distinct neural networks, underlying each of two differentiated outcomes. Then one set of synapses must be strengthened and the other allowed to decay. What fires apart wires apart.

Empowerment is an explicit aim of treatment in nonmedical approaches to addiction. For example, SMART Recovery is a therapeutic approach integrating cognitive behavioural methods with interpersonal support in group settings. This organization encourages self-empowerment and self-directed change, and its facilitators teach methods for grooming those capacities. Motivational interviewing techniques (which have also been incorporated into SMART) are especially designed to harness the client's motivation and direct it toward self-selected goals in an atmosphere of compassion and support. However, in these and other current approaches, motivation is accessed by cognitive means. Clients are urged to explore their choices, evaluate likely outcomes, compare those outcomes, and decide what they will pursue. Or they may be encouraged to examine and modify their own belief systems. Cognition-based techniques can certainly be helpful, but they lose potency when they overlook the visceral reality of moment-to-moment desire. As we have seen, what looks good at one point in time might be meaningless or irrelevant a moment later. That's why desire has to be connected with choice, and that can only take place in the moment.

The only way to forge new connections between the motivational core and the dorsolateral PFC is to activate both at the same time. What fires together wires together. Desire has to ignite the striatum, while judgement and perspective are held in place by the PFC. This formula reminds me of a snake swallowing its tail: our brains can envision a future that extends almost limitlessly, yet our brains can only be accessed in the moment, because their chemical and structural composition changes so quickly. If addicts can access self-narrative in the heat of the moment, even briefly, then desire has the potential to jump like static electricity to new possibilities and burnish them with drive. That can be a difficult juggling act. Yet, while there is no single formula for making it work, there are several ways to give the juggler the best possible chance.

HELPING PEOPLE QUIT

How does a close look at now appeal, miscommunicating brain structures, and self-narrative add to our repertoire of tactics for overcoming addiction? You don't need neuroscience to argue for the role of empowerment in addiction treatment. And people often overcome their addictions without any treatment at all. Yet quitting will be facilitated, either in or out of treatment, if we take our brains seriously—if we recognize the power of feelings, the chemical urgency with which they arise, their mercurial adjustments to time and circumstances, and their exquisite capacity to focus attention. Brain science helps us to understand how desire works and how it connects us with goals—new goals as well as old ones. In my view, any recipe for change will be enhanced by this knowledge.

To see what this approach can offer addicts trying to quit, let's start with a look at the failures of medicalization—failures protracted by the broad acceptance of the disease model. As noted at the beginning of this chapter, the medicalization of addiction has provided certain benefits. Foremost among them has been the development of pharmaceutical agents that can diminish withdrawal symptoms and ease cravings. Even if these are temporary measures, they can make a real difference during the darkest of times. The disease model has also led medicine and society to a more enlightened view of addiction, as a very human phenomenon with clear biological underpinnings, while encouraging humane treatment for those who suffer. But treatment approaches based on the disease model are too often ineffective. Addicts remain addicted despite their doctors' best efforts. Medicines that help people cope with symptoms do not ignite the desire to change or light up new pathways for life beyond addiction. Worse still, rigid, cookie-cutter methods and institutional self-interests too often turn "treatment" into a dead end or a revolving door for people who seek help. The premise of this book is that medicalization and the disease model have outlived their usefulness.

To bring the drawbacks of medicalization to a point, consider my claim that addiction can only be beaten by the alignment of desire with personally derived, future-oriented goals. Does medically based treatment help with that agenda? On the contrary, such treatment is almost always institutional treatment, and institutions are famous for eroding the self-direction that addicts may have mustered to get them to the door. Typically, those seeking treatment are told to call back, unless they are ordered into treatment by the legal system, which obviously trounces self-direction from the outset. Then they are given a date to come in for an assessment. Any delay can be easily justified: "We want to make sure you're really ready." Finally they're scheduled to begin treatment, weeks later. That is, if they're lucky enough to bypass the notoriously long waiting lists for state-sponsored care or afford the swank offerings of a private setting. They are assigned a bed. Ironically, their beds are the hallmark of their claim for help, but beds are where people sleep and where sick people lie when they can't walk around; they are hardly platforms for initiative and empowerment. Then, if the waiting time for service delivery hasn't completely undermined their incentive to change, the philosophy of medical care may do so. Addicts become patients, and patients do not participate in decisions about their care. Patients follow the regimens handed down by authority figures who understand the workings of their disease far better than they do. So personal intention has no place in the cure.

If you think this depiction is too extreme, you need only listen to the stories of addicts who have been through institutional care, or read the striking accounts compiled by Anne Fletcher in *Inside Rehab*.[4] Even addicts determined to quit often feel overwhelmed by the weight of depersonalization, passivity, and submission to authority, the disinterest of staff in their personal views, and their exclusion from evaluations of how they're doing, what they're doing, and when they've had enough. At the outset they are told, "We'll have to break you down so we can build you back up again"—a phrase commonly heard in institutional settings, according to Matt Robert, a friend, former addict, and group facilitator in both institutional and community-based programs. It's not that such policies are born of ill intent. It's just that they're wrong-headed. Disease model advocates like David Sack (introduced in Chapter One) despair that "a large portion of addicts continue to use in the years following treatment regardless of the particular drug involved."[5] They view this as evidence that the disease of addiction is terribly serious and needs all the ammunition society can muster—which often translates to more money and more institutional beds. Yet the obvious conclusion is that mainstream treatment for addiction just doesn't work. And since it is founded on the disease model, that model is probably flawed.

What alternatives might stem from a developmental approach to treatment that applies the power of momentary desire to a personal timeline for quitting? Most important, no single strategy, organization, method, or philosophy commands centre stage. Any initiative that meets addicts when and where they're ready to quit is well positioned to help them move onward. Community-based settings can fill this role most easily, because there is no fortress wall that needs to be scaled, no lineup at the door, and no financial minefield that needs to be crossed. Nor are there likely to be rigid policies that preempt the addict's personal incentive. When desire is ready

to arc from the goal of immediate relief to the goal of a valued future, treatment can begin. Not by inducing desire—only frustration
and suffering can do that—but by capturing and holding one's vision of that future. I've compared group processes in recovery to
the societal narratives that helped Native youth in Canada maintain
a personal story and imagine a viable future. In community-based
groups, including AA and SMART Recovery, this kind of narrative
scaffolding is ready and waiting for addicts who are ready to quit
(though other important resources may be lacking). Group meetings are frequently inserted into institutional treatment as well, but
whether they're available when addicts are ready to take the plunge
is entirely hit-and-miss. And while group processes can be helpful,
they are certainly not always helpful, nor are they the only way forward. A therapist who is ready to start precisely when the addict is
ready to stop can be immensely valuable. Treatment just requires
the attention of one other human being who can hold, possibly
distil, and hopefully extend the narrative energized by the cresting
wave of an individual's desire to change.

What will work best is whatever is available when the synaptic
avenues of desire make contact with brain regions responsible for
perspective change. This can be the presence of a friend who accompanies you to your first twelve-step meeting, as was the case for
Brian at the very point when he'd had enough. Or the attention of
a therapist who really gets where you've been and where you want
to go, as was the case for Donna at the fulcrum of her despair. It
can even be the horrific embrace of a jail cell where you see your
options with brutal clarity, as was true for Natalie. It can be a month
on your grandfather's farm, a book that captures your heart when
you think you've lost it, or the opening of a long-stuck window
through meditation, romance, or antidepressant therapy.

Quitting requires a merger, perhaps a collision, between desire
and perspective—again, what fires together wires together—yet it

doesn't demand any particular brand of intervention. Nevertheless, I'll end by sketching a radical treatment initiative that I think exemplifies a real-time approach to battling addiction, launched by my friend Peter Sheath. Peter is another former addict and senior associate of a consulting group for service delivery in the United Kingdom. Under the auspices of Reach Out Recovery (ROR), the city of Birmingham is embarking on a campaign to be there for addicts at the very moment when their desire for change is ignited. The program's intent is to distribute treatment nodes across community sites that are most available to addicts in their day-to-day lives. Shopkeepers, including newsagents, bakers, butchers, and pharmacists, are trained in brief interventions. Their "recovery-friendly" shops display an ROR sticker on the front window, so addicts are aware that they can go there for help. People come in off the street, perhaps buying a loaf of bread at the same time, and say, "I've had enough! I'm ready to quit!" Then the shopkeeper tells them they've come to the right place, takes a quick inventory, and advises them on what to do next. Perhaps "Hey, don't drink too much today, then come on back and talk with me at five o'clock." Or, in more severe cases, "I won't be able to work with you, but I know somebody who can." People might be referred to peer mentors who will show up the following day to help them with difficult issues such as detox and family matters. Participating doctors are on hand for those who need medical care. And phone numbers for immediate telehealth support are posted in plain view, to help initiate and coordinate services. Even taxi drivers have been recruited, so that someone en route to score can throw up his hands and call it quits without letting that incentive get lost along the way.

This strikes me as a highly creative approach, brimming with possibility. Whether it will work as well as we hope remains to be seen. The so-called Birmingham Model is just one example of an approach that resonates with the arguments I've made in this chapter.

But I've described it in detail also because it exemplifies the innovations waiting to emerge when the disease model is retracted and a fresh perspective, free of orthodoxy and special interests, is allowed to take its place. The Birmingham Model was inspired by intuitions about the mercurial nature of desire and the critical role of timing in addiction treatment. But it also rides on the insight that addicts aren't diseased and they don't need medical intervention in order to change their lives. What they need is sensitive, intelligent social scaffolding to hold the pieces of their imagined future in place—while they reach toward it.

∾

If the brain region that allows us to imagine the future is synched up with the brain regions that propel us toward our goals, and if that linkage is practiced and reinforced, so that synaptic highways become smooth and efficient, then addiction need be no more than a stage in the development of the self. And that often seems to be exactly what it is. Despite the misery they may have experienced, quite a few former addicts have told me that they wouldn't be who they are now without the struggles they endured while trying to quit. As a neuroscientist, I view this passage the way a city planner might recall the construction of an overpass to relieve snarled traffic. As a developmentalist, I see it as a vivid instance of the role of suffering in individual growth. And as someone who has known addiction personally, I recognize it as the bounce our lives can take when they've hit bottom once too often.

ACKNOWLEDGEMENTS

After writing a book about my own passage through addiction, I needed to learn what my experiences had in common with those of others. So I began a regular blog that attracted a bright, boisterous, and empathic community populated by former and recovering addicts. The many comments following my posts and the guest posts contributed by members provided a wealth of insights and information that I could not have hoped to find elsewhere. I want to thank each and every one of the people who've engaged in this conversation with me. You inspired me to write the present book, and you helped me understand addiction well enough to feel I could make a worthwhile contribution.

The five former addicts whose stories I tell deserve the gratitude of everyone attempting to comprehend addiction by combining private experience with other forms of knowledge. The people who volunteered for this project donated many hours to respond to my questions, and they did so with unstinting energy and honesty, dredging up details from experiences they might have preferred to forget. When wearing my interviewer's hat, I often felt like a dentist drilling deeply, painfully, until I unearthed every chunk of my respondent's past. They bore up bravely, shining the beam of self-examination wherever I asked them to look. I am deeply grateful.

Lisa Kaufman, my editor at PublicAffairs, helped me upgrade my understanding of the rehab world, past and present, until I'd acquired the perspective I needed to portray it sensitively and accurately. But I'm most grateful to Lisa for encouraging me to follow the implications of my own model from theoretical abstractions to concrete directions for practice. She convinced me that, for many readers, that's where the book had to land. And she was right.

Tim Rostron, my editor at Doubleday Canada, has now been my writing guru through two books, and I continue to celebrate my good fortune. Tim's mastery of the deep and subtle currents of English and his dedication to transparency and flow have nursed my growth from scientist to writer.

I benefited hugely from the seasoned perspective of two unpaid editors, Matt Robert and Cathy O'Connor. As a pioneer in the rehab community and a sparkling commentator on current trends, Matt took me behind the scenes of the rehab/recovery world. He read most if not all of these chapters, showed me what I was missing in both form and substance, and helped me smooth out terms and concepts that might otherwise get caught in the reader's throat. Cathy generously dipped into her editorial talents to guide me through the no-man's-land between what I thought I was explaining clearly and what readers were likely to grasp. There were jagged craters everywhere, most in places I would not have checked. Cathy pointed them out with patience and precision and helped me figure out how to fill them. I am so very grateful to both of you.

Other treatment experts came to my aid. I am particularly indebted to Shaun Shelly, who kept pace with every conceptual step I took, in the book and in the blog, and harvested examples to help support our shared understanding of addiction. And my thanks to Peter Sheath, who spearheaded the Birmingham Model described in the last chapter and infected me with the courage, creativity, and optimism he has brought to the treatment world.

My most generous and dependable editor remains Isabela Granic, my partner for eighteen years. Your steady supply of gist was the mortar that allowed my details to cohere and settle. You continued pointing me toward what I'm good at and reminding me of its worth. And you stoked the fires whenever I got discouraged or just tired. This book could not have existed without you.

Finally, Ruben and Julian, thank you for letting me work all those hours when I should have been playing with you. Ruben, thanks for adjusting my chair. Julian, thanks for the cuddles. I'll try to make it up to you both now that the book is finished.

NOTES

Chapter 1: Defining Addiction

1. K. T. Brady and R. Sinha, "Co-occurring Mental and Substance Use Disorders: The Neurobiological Effects of Chronic Stress," *American Journal of Psychiatry* 162, no. 8 (2005): 1483–1493; Mustafa al'Absi, *Stress in Addiction: Biological and Psychological Mechanisms* (Amsterdam: Academic Press, 2006).

2. The prominence of the disease model is discussed and debated in Matt Field, "Addiction Is a Brain Disease...but Does It Matter?," *The Mental Elf*, February 6, 2015, www.thementalelf.net/mental-health-conditions/substance-misuse/addiction-is-a-brain-diseasebut-does-it-matter.

3. National Institute on Drug Abuse, "The Science of Drug Abuse and Addiction: The Basics," September 2014, www.drugabuse.gov/publications/media-guide/science-drug-abuse-addiction.

4. Interview of Steven Hyman by Bill Moyers, WNET website, www.thirteen.org/closetohome/science/html/hyman.html.

5. Nora D. Volkow, "Preface: How Science Has Revolutionized the Understanding of Drug Addiction," in National Institute on Drug Abuse, *Drugs, Brains, and Behavior: The Science of Addiction* (Washington, DC: NIDA, 2014).

6. National Institute on Drug Abuse, *Drugs, Brains, and Behavior: The Science of Addiction* (Washington, DC: NIDA, 2014), 5.

7. Data on child poverty from Max Fisher, "Map: How 35 Countries Compare on Child Poverty (the U.S. Is Ranked 34th)," *WorldViews* blog, *Washington Post*, April 15, 2013, www.washingtonpost.com/blogs/worldviews/wp/2013/04/15/map-how-35-countries-compare-on-child-poverty-the-u-s-is-ranked-34th.

8. William R. Miller, Verner S. Westerberg, Richard J. Harris, and J. Scott Tonigan, "What Predicts Relapse? Prospective Testing of Antecedent Models," *Addiction* 91 supp. (December 1996): S155–171.

9. For examples, see data from the National Epidemiologic Survey on Alcohol and Related Conditions, summarized in the National Institute on Alcohol Abuse and Alcoholism's *Alcohol Alert* 70 (October 2006), http://pubs.niaaa.nih.gov/publications/AA70/AA70.htm, and a comprehensive review and analysis by Gene Heyman, "Quitting Drugs: Quantitative and Qualitative Features," *Annual Review of Clinical Psychology* 9 (2013): 29–59.

10. Note how this phrasing echoes NIDA's content and tone. And thanks to Stanton Peele and his blog for *Psychology Today* (www.psychologytoday.com/blog/addiction-in-society) for bringing this incredible moment in US politics to our attention.

11. See the comprehensive "Combined Addiction Disease Chronologies" by William White, Ernest Kurtz, and Caroline Acker at www.silkworth.net/kurtz.

12. From George E. Pettey, *The Narcotic Drug Diseases and Allied Ailments* (Philadelphia: F. A. Davis, 1913), 5–6, 192; thanks to White, Kurtz, and Acker, "Combined Addiction Disease Chronologies."

13. Alan I. Leshner summarized the brain disease model for the scientific community in a flagship article, "Addiction Is a Brain Disease, and It Matters," *Science* 278, no. 5335 (1997): 45–47.

14. Jeffrey M. Jones, "Americans with Addiction in Their Family Believe It Is a Disease," Gallup News Service, August 11, 2006, www.gallup.com/poll/24097/americans-addiction-their-family-believe-disease.aspx.

15. David Sack, "Addiction Is a Disease and Needs to Be Treated as Such," *Room for Debate* blog, *New York Times*, February 11, 2014, www.nytimes.com/roomfordebate/2014/02/10/what-is-addiction/addiction-is-a-disease-and-needs-to-be-treated-as-such. Sack's personal website is www.drdavidsack.com.

16. S. Darke et al., "Patterns of Sustained Heroin Abstinence Amongst Long-Term, Dependent Heroin Users: 36 Months Findings from the Australian Treatment Outcome Study (ATOS)," *Addictive Behavior* 32, no. 9 (2007): 1897–1906.

17. Dirk Hanson, "Ivan Oransky on the Disease Model at TEDMED 2012," *Addiction Inbox: The Science of Substance Abuse*, April 15, 2012, http://addiction-dirkh.blogspot.com.es/2012/04/ivan-oransky-on-disease-model-at-tedmed.html.

18. Gene Heyman, *Addiction: A Disorder of Choice* (Cambridge, MA: Harvard University Press, 2009).

19. Heyman, "Quitting Drugs."

20. An easily accessible review of these studies is Bruce K. Alexander, "Addiction: The View from Rat Park," www.brucekalexander.com/articles-speeches/rat-park/148-addiction-the-view-from-rat-park.

21. Carl Hart, *High Price: Drugs, Neuroscience, and Discovering Myself* (New York: Harper, 2013).

22. Stanton Peele and Ilse Thompson, *Recover! Stop Thinking Like an Addict and Reclaim Your Life with The PERFECT Program* (New York: Da Capo Press, 2014).

23. S. Satel and S. O. Lilienfeld, "Addiction and the Brain-Disease Fallacy," *Frontiers in Psychiatry* 4 (2013): 141.

24. Italics added. P. Rakic, J.-P. Bourgeois, and P. S. Goldman-Rakic, "Synaptic Development of the Cerebral Cortex: Implications for Learning, Memory, and Mental Illness," in J. van Pelt, M. A. Corner, H. B. M. Uylings, and F. H. Lopes da Silva, eds., *The Self-Organizing Brain: From Growth Cones to Functional Networks*, Progress in Brain Research vol. 102 (Amsterdam: Elsevier, 1994), 227–243.

Chapter 2: A Brain Designed for Addiction

1. From William James's *Habit* (1887), quoted by Maria Popova at www.brain pickings.org/index.php/2012/09/25/william-james-on-habit.

Chapter 3: When Craving Comes to Power

(There are no notes for this chapter.)

Chapter 4: The Tunnel of Attention

1. For a meta-analysis, see J. MacKillop, M. T. Amlung, L. R. Few, L. A. Ray, L. H. Sweet, and M. R. Munafo, "Delayed Reward Discounting and Addictive Behavior: A Meta-analysis," *Psychopharmacology* 216 (2011): 305–321.

Chapter 5: Donna's Secret Identity

(There are no notes for this chapter.)

Chapter 6: Johnny Needs a Drink

1. Barry J. Everitt and Trevor W. Robbins, "From the Ventral to the Dorsal Striatum: Devolving Views of Their Role in Drug Addiction," *Neuroscience and Biobehavioral Reviews* 37, no. 9, part A (2013): 1946–1954.

2. See Rita Z. Goldstein and Nora D. Volkow, "Dysfunction of the Prefrontal Cortex in Addiction: Neuroimaging Findings and Clinical Implications," *Nature Reviews Neuroscience* 12 (2011): 652–669. But note that these findings may mean something else completely. They may mean that people who already have less grey matter volume in these prefrontal areas are more likely to take drugs for longer periods of time. Present-day study designs are incapable of determining which interpretation is valid, or whether both have credence.

3. C. G. Connolly, R. P. Bell, J. J. Foxe, and H. Garavan, "Dissociated Grey Matter Changes with Prolonged Addiction and Extended Abstinence in Cocaine Users," *PLOS ONE* 8, no. 3 (2013): e59645.

4. We must evaluate these findings cautiously, as they are based on a single study, but they are certainly interesting, and Hugh Garavan's interpretation is similar to mine (personal communication, September 2014).

Chapter 7: Nothing for Alice

1. Barry J. Everitt and Trevor W. Robbins, "From the Ventral to the Dorsal Striatum: Devolving Views of Their Role in Drug Addiction," *Neuroscience and Biobehavioral Reviews* 37, no. 9, part A (2013): 1946–1954.

Chapter 8: Biology, Biography, and Addiction

1. See Helen E. Fisher, Arthur Aron, Debra Mashek, Haifang Li, and Lucy L. Brown, "Defining the Brain Systems of Lust, Romantic Attraction, and Attachment," *Archives of Sexual Behavior* 31, no. 5 (2002): 413–419.

2. James P. Burkett and Larry J. Young, "The Behavioral, Anatomical and Pharmacological Parallels Between Social Attachment, Love, and Addiction," *Psychopharmacology* 224, no. 1 (2012): 1–26. For ease of understanding, I have replaced acronyms for anatomical structures with full names.

3. Ibid.

4. Gabor Maté, in his book *In the Realm of Hungry Ghosts* (Toronto: Vintage Canada, 2008), has documented the links among early adversity, alterations in neural development, and addiction in later life.

5. Maia Szalavitz, "How Childhood Trauma Creates Life-long Adult Addicts," *The Fix*, September 25, 2011.

6. Nikolaus Steinbeis, Johannes Haushofer, Ernst Fehr, and Tania Singer, "Development of Behavioral Control and Associated vmPFC–DLPFC Connectivity Explains Children's Increased Resistance to Temptation in Intertemporal Choice," *Cerebral Cortex*, 2014; doi:10.1093/cercor/bhu167.

7. B. Figner, D. Knoch, E. J. Johnson, A. R. Krosch, S. H. Lisanby, E. Fehr, and E. U. Weber, "Lateral Prefrontal Cortex and Self-Control in Intertemporal Choice," *Nature Neuroscience* 13, no. 5 (2010): 538–539; doi:10.1038/nn.2516.

8. See Xing Tian and David E. Huber, "Playing 'Duck Duck Goose' with Neurons: Change Detection Through Connectivity Reduction," *Psychological Science* 24, no. 6 (2013): 819–827.

Chapter 9: Developing Beyond Addiction

1. A. Mogilner, J. A. Grossman, U. Ribary, M. Joliot, J. Volkmann, D. Rapoport, R. W. Beasley, and R. R. Llinás, "Somatosensory Cortical Plasticity in Adult Humans Revealed by Magnetoencephalography," *Proceedings of the National Academy of Sciences* 90, no. 8 (1993): 3593–3597.

2. See Michael J. Chandler and Christopher Lalonde, "Cultural Continuity as a Hedge Against Suicide in Canada's First Nations," *Transcultural Psychiatry* 35, no. 2 (1998): 191–219.

3. Bruce Alexander, *The Globalization of Addiction* (Oxford: Oxford University Press, 2008).

4. Anne M. Fletcher, *Inside Rehab: The Surprising Truth About Addiction Treatment—and How to Get Help That Works* (New York: Viking, 2013).

5. David Sack, "Addiction Is a Disease and Needs to Be Treated as Such," *Room for Debate* blog, *New York Times,* February 11, 2014.

INDEX

TMS. *See* transcranial magnetic
   stimulation
traditions, 204
transcranial magnetic stimulation
   (TMS), 186, 187 (fig.)
trauma, 3
twelve-step programs (groups,
   meetings, philosophy)
  abstinence and, 15
  alternative resources and, 15
  character and, 15
  disease terminology in, 14
  institutional treatment and, 14–15
  powerlessness admission in, 13
  See also Alcoholics Anonymous;
    Minnesota Model; Narcotics
    Anonymous

ventral striatum, 44 (fig.), 45

impulsive action and, 125–126
impulsive to compulsive behaviour
   and, 127–128
See also accumbens
Vicodin, 100, 101
Vietnam veterans, 3, 21
Volkow, Nora
  binge eating and, 146, 150
  disease model and, 7, 8–9
  learning and, 8

webbed fingers, 195–196, 196 (fig.)
White, William, 21
will, 191–192
Wilson, Bill, 12
The Woman Who Changed Her Brain
   (Arrowsmith Young), 194–195

Young, Larry, 166–167

Courtesy of the author

Dr. Marc Lewis is a neuroscientist and professor of developmental psychology, now teaching at Radboud University in the Netherlands after more than twenty years on faculty at the University of Toronto. He has authored or co-authored more than fifty journal articles in neuroscience and developmental psychology. Presently, he speaks and blogs on topics in addiction science, and his critically acclaimed book, *Memoirs of an Addicted Brain: A Neuroscientist Examines His Former Life on Drugs,* is the first to blend memoir and science in addiction studies.

PublicAffairs is a publishing house founded in 1997. It is a tribute to the standards, values, and flair of three persons who have served as mentors to countless reporters, writers, editors, and book people of all kinds, including me.

I. F. STONE, proprietor of *I. F. Stone's Weekly*, combined a commitment to the First Amendment with entrepreneurial zeal and reporting skill and became one of the great independent journalists in American history. At the age of eighty, Izzy published *The Trial of Socrates*, which was a national bestseller. He wrote the book after he taught himself ancient Greek.

BENJAMIN C. BRADLEE was for nearly thirty years the charismatic editorial leader of *The Washington Post*. It was Ben who gave the *Post* the range and courage to pursue such historic issues as Watergate. He supported his reporters with a tenacity that made them fearless and it is no accident that so many became authors of influential, best-selling books.

ROBERT L. BERNSTEIN, the chief executive of Random House for more than a quarter century, guided one of the nation's premier publishing houses. Bob was personally responsible for many books of political dissent and argument that challenged tyranny around the globe. He is also the founder and longtime chair of Human Rights Watch, one of the most respected human rights organizations in the world.

. . .

For fifty years, the banner of Public Affairs Press was carried by its owner Morris B. Schnapper, who published Gandhi, Nasser, Toynbee, Truman, and about 1,500 other authors. In 1983, Schnapper was described by *The Washington Post* as "a redoubtable gadfly." His legacy will endure in the books to come.

Peter Osnos, *Founder and Editor-at-Large*